EMERGENCE

Also by Steven Johnson

Interface Culture:
How New Technology Transforms the Way We Create and Communicate

EMERGENCE

THE CONNECTED LIVES OF
ANTS, BRAINS, CITIES AND SOFTWARE

STEVEN JOHNSON

ALLEN LANE
THE PENGUIN PRESS

ALLEN LANE
THE PENGUIN PRESS

Published by the Penguin Group
Penguin Books Ltd, 80 Strand, London WC2R 0RL, England
Penguin Putnam Inc., 375 Hudson Street, New York, New York 10014, USA
Penguin Books Australia Ltd, 250 Camberwell Road, Camberwell, Victoria 3124, Australia
Penguin Books Canada Ltd, 10 Alcorn Avenue, Toronto, Ontario, Canada M4V 3B2
Penguin Books India (P) Ltd, 11, Community Centre, Panchsheel Park, New Delhi – 110 017, India
Penguin Books (NZ) Ltd, Cnr Rosedale and Airborne Roads, Albany, Auckland, New Zealand
Penguin Books (South Africa) (Pty) Ltd, 24 Sturdee Avenue, Rosebank 2196, South Africa

Penguin Books Ltd, Registered Offices: 80 Strand, London WC2R 0RL, England

www.penguin.com

First published in the USA by Scribner 2001
First published in Great Britain by Allen Lane The Penguin Press 2001
2

Printed and bound in Great Britain by Clays Ltd, St Ives plc
Cover repro and printing by Concise Cover Printers

for my wife

CONTENTS

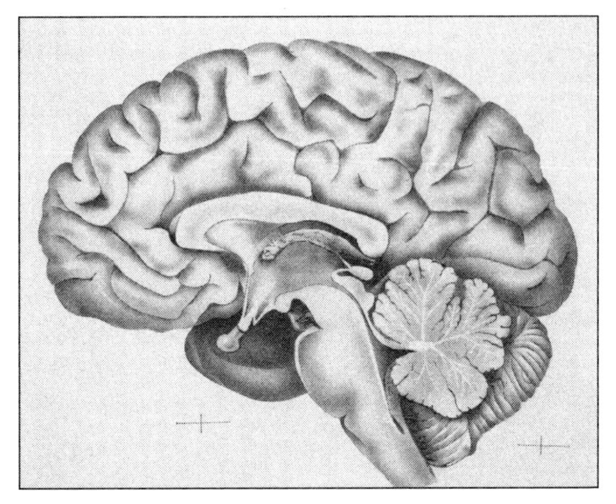

Diagram of the human brain *(Courtesy of Mittermeier)*

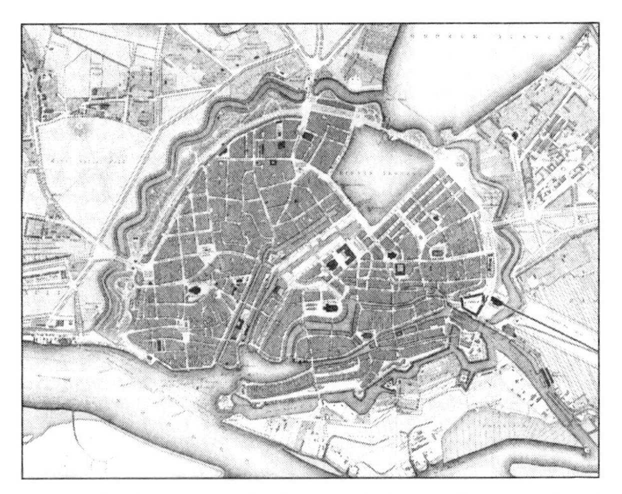

Map of Hamburg, circa 1850 *(Courtesy of Princeton Architectural Press)*

Most of all, we need to preserve the absolute unpredictability and total improbability of our connected minds. That way we can keep open all the options, as we have in the past.

It would be nice to have better ways of monitoring what we're up to so that we could recognize change while it is occurring.... Maybe computers can be used to help in this, although I rather doubt it. You can make simulation models of cities, but what you learn is that they seem to be beyond the reach of intelligent analysis. . . . This is interesting, since a city is the most concentrated aggregation of humans, all exerting whatever influence they can bring to bear. The city seems to have a life of its own. If we cannot understand how this works, we are not likely to get very far with human society at large.

Still, you'd think there would be some way in. Joined together, the great mass of human minds around the earth seems to behave like a coherent, living system. The trouble is that the flow of information is mostly one-way. We are all obsessed by the need to feed information in, as fast as we can, but we lack sensing mechanisms for getting anything much back. I will confess that I have no more sense of what goes on in the mind of mankind than I have for the mind of an ant. Come to think of it, this might be a good place to start.

—Lewis Thomas, 1973

INTRODUCTION

Here Comes Everybody!

In August of 2000, a Japanese scientist named Toshiyuki Nakagaki announced that he had trained an amoebalike organism called slime mold to find the shortest route through a maze. Nakagaki had placed the mold in a small maze comprising four possible routes and planted pieces of food at two of the exits. Despite its being an incredibly primitive organism (a close relative of ordinary fungi) with no centralized brain whatsoever, the slime mold managed to plot the most efficient route to the food, stretching its body through the maze so that it connected directly to the two food sources. Without any apparent cognitive resources, the slime mold had "solved" the maze puzzle.

For such a simple organism, the slime mold has an impressive intellectual pedigree. Nakagaki's announcement was only the latest in a long chain of investigations into the subtleties of slime mold behavior. For scientists trying to understand systems that use relatively simple components to build higher-level intelligence, the slime mold may someday be seen as the equivalent of the finches and tortoises that Darwin observed on the Galápagos Islands.

How did such a lowly organism come to play such an important scientific role? That story begins in the late sixties in New York City, with a scientist named Evelyn Fox Keller. A Harvard Ph.D. in physics, Keller had written her dissertation on molecular biology, and she had spent some time exploring the nascent field of "non-equilibrium thermodynamics," which in later years would come to be associated with complexity theory. By 1968, she was working as an associate at Sloan-Kettering in Manhattan, thinking about the application of mathematics to biological problems. Mathematics had played such a tremendous role in expanding our understanding of physics, Keller thought—so perhaps it might also be useful for understanding living systems.

In the spring of 1968, Keller met a visiting scholar named Lee Segel, an applied mathematician who shared her interests. It was Segel who first introduced her to the bizarre conduct of the slime mold, and together they began a series of investigations that would help transform not just our understanding of biological development but also the disparate worlds of brain science, software design, and urban studies.

If you're reading these words during the summer in a suburban or rural part of the world, chances are somewhere near you a slime mold is growing. Walk through a normally cool, damp section of a forest on a dry and sunny day, or sift through the bark mulch that lies on a garden floor, and you may find a grotesque substance coating a few inches of rotting wood. On first inspection, the reddish orange mass suggests that the neighbor's dog has eaten something disagreeable, but if you observe the slime mold over several days—or, even better, capture it with time-lapse photography—you'll discover that it moves, ever so slowly, across the soil. If the weather conditions grow wetter and cooler, you may return to the same spot and find the creature has disappeared altogether. Has it wandered off to some other part of the forest? Or somehow vanished into thin air, like a puddle of water evaporating?

As it turns out, the slime mold *(Dictyostelium discoideum)* has done something far more mysterious, a trick of biology that had confounded scientists for centuries, before Keller and Segel began their collaboration. The slime mold behavior was so odd, in fact, that understanding it required thinking outside the boundaries of traditional disciplines—which may be why it took a molecular biologist with a physics Ph.D.'s instincts to unravel the slime mold's mystery. For that is no disappearing act on the garden floor. The slime mold spends much of its life as thousands of distinct single-celled units, each moving separately from its other comrades. Under the right conditions, those myriad cells will coalesce again into a single, larger organism, which then begins its leisurely crawl across the garden floor, consuming rotting leaves and wood as it moves about. When the environment is less hospitable, the slime mold acts as a single organism; when the weather turns cooler and the mold enjoys a large food supply, "it" becomes a "they." The slime mold oscillates between being a single creature and a swarm.

While slime mold cells are relatively simple, they have attracted a disproportionate amount of attention from a number of different disciplines—embryology, mathematics, computer science—because they display such an intriguing example of coordinated group behavior. Anyone who has ever contemplated the great mystery of human physiology—how do all my cells manage to work so well together?—will find something resonant in the slime mold's swarm. If we could only figure out how the *Dictyostelium* pull it off, maybe we would gain some insight on our own baffling togetherness.

"I was at Sloan-Kettering in the biomath department—and it was a very small department," Keller says today, laughing. While the field of mathematical biology was relatively new in the late sixties, it had a fascinating, if enigmatic, precedent in a then-little-known essay written by Alan Turing, the brilliant English code-breaker from World War II who also helped invent the digital computer. One of Turing's last published papers, before his death in 1954, had studied the riddle of "morphogenesis"—the capacity of all life-forms to develop ever more baroque bodies out of impossibly simple beginnings. Turing's paper had focused more on the recurring numerical patterns of flowers, but it demonstrated using mathematical tools how a complex organism could assemble itself without any master planner calling the shots.

"I was thinking about slime mold aggregation as a model for thinking about development, and I came across Turing's paper," Keller says now, from her office at MIT. "And I thought: Bingo!"

For some time, researchers had understood that slime cells emitted a common substance called acrasin (also known as cyclic AMP), which was somehow involved in the aggregation process. But until Keller began her investigations, the conventional belief had been that slime mold swarms formed at the command of "pacemaker" cells that ordered the other cells to begin aggregating. In 1962, Harvard's B. M. Shafer showed how the pacemakers could use cyclic AMP as a signal of sorts to rally the troops; the slime mold generals would release the compounds at the appropriate moments, triggering waves of cyclic AMP that washed through the entire community, as each isolated cell relayed the signal to its neighbors. Slime mold aggregation, in effect, was a giant game of Telephone—but only a few elite cells placed the original call.

It seemed like a perfectly reasonable explanation. We're naturally predisposed to think in terms of pacemakers, whether we're talking about fungi, political systems, or our own bodies. Our actions seem governed for the most part by the pacemaker cells in our brains, and for millennia we've built elaborate pacemakers cells into our social organizations, whether they come in the form of kings, dictators, or city councilmen. Much of the world around us can be explained in terms of command systems and hierarchies—why should it be any different for the slime molds?

But Shafer's theory had one small problem: no one could find the pacemakers. While all observers agreed that waves of cyclic AMP did indeed flow through the slime mold community before aggregation, all the cells in the community were effectively interchangeable. None of them possessed any distinguishing characteristics that might elevate them to pacemaker status. Shafer's theory had presumed the existence of a cellular monarchy commanding the masses, but as it turned out, all slime mold cells were created equal.

For the twenty years that followed the publication of Shafer's original essay, mycologists assumed that the missing pacemaker cells were a sign of insufficient data, or poorly designed experiments: The generals were there somewhere in the mix, the scholars assumed—they just didn't know what their uniforms looked like yet. But Keller and Segel took another, more radical approach. Turing's work on morphogenesis had sketched out a mathematical model wherein simple agents following simple rules could generate amazingly complex structures; perhaps the aggregations of slime mold cells were a real-world example of that behavior. Turing had focused primarily on the interactions between cells in a single organism, but it was perfectly reasonable to assume that the math would work for aggregations of free-floating cells. And so Keller started to think: What if Shafer had it wrong all along? What if the community of slime mold cells were organizing themselves? What if there were no pacemakers?

Keller and Segel's hunch paid off dramatically. While they lacked the advanced visualization tools of today's computers, the two scratched out a series of equations using pen and paper, equations that demonstrated how slime cells could trigger aggregation without following a leader, simply by altering the amount of cyclic AMP they released individually, then following trails of the pheromone that they encountered as they wandered through their environment. If the slime cells pumped out enough cyclic AMP, clusters of cells would start to form. Cells would begin following trails created by other cells, creating a positive feedback loop that encouraged more cells to join the cluster. If each solo cell was simply releasing cyclic AMP based on its own local assessment of the general conditions, Keller and Segel argued in a paper published in 1969, then the larger slime mold community might well be able to aggregate based on global changes in the environment—all without a pacemaker cell calling the shots.

"The response was very interesting," Keller says now. "For anyone who understood applied mathematics, or had any experience in fluid dynamics, this was old hat to them. But to biologists, it didn't make any sense. I would give seminars to biologists, and they'd say, 'So? Where's the founder cell? Where's the pacemaker?' It didn't provide any satisfaction to them whatsoever." Indeed, the pacemaker hypothesis would continue as the reigning model for another decade, until a series of experiments convincingly proved that the slime mold cells were organizing from below. "It amazes me how difficult it is for people to think in terms of collective phenomenon," Keller says today.

Thirty years after the two researchers first sketched out their theory on paper, slime mold aggregation is now recognized as a classic case study in bottom-up behavior. Keller's colleague at MIT Mitch Resnick has even developed a computer simulation of slime mold cells aggregating, allowing students to explore the eerie, invisible hand of self-organization by altering the number of cells in the environment, and the levels of cyclic AMP distributed. First-time users of Resnick's simulation invariably say that the on-screen images—brilliant clusters of red cells and green pheromone trails—remind them of video games, and in fact the comparison reveals a secret lineage. Some of today's most popular computer games resemble slime mold cells because they are loosely based on the equations that Keller and Segel formulated by hand in the late sixties. We like to talk about life on earth evolving out of the primordial soup. We could just as easily say that the most interesting digital life on our computer screens today evolved out of the slime mold.

You can think of Segel and Keller's breakthrough as one of the first few stones to start tumbling at the outset of a landslide. Other stones were moving along with theirs—some of whose trajectories we'll follow in the coming pages—but that initial movement was nothing compared to the avalanche that followed over the next two decades. At the end of its course, that landslide had somehow conjured up a handful of fully credited scientific disciplines, a global network of research labs and think tanks, and an entire patois of buzzwords. Thirty years after Keller challenged the pacemaker hypothesis, students now take courses in "self-organization studies," and bottom-up software helps organize the Web's most lively virtual communities. But Keller's challenge did more than help trigger a series of intellectual trends. It also unearthed a secret history of decentralized thinking, a history that had been submerged for many years beneath the weight of the pacemaker hypothesis and the traditional boundaries of scientific research. People had been thinking about emergent behavior in all its diverse guises for centuries, if not millennia, but all that thinking had consistently been ignored as a unified body of work—because there was nothing unified about its body. There were isolated cells pursuing the mysteries of emergence, but no aggregation.

Indeed, some of the great minds of the last few centuries—Adam Smith, Friedrich Engels, Charles Darwin, Alan Turing—contributed to the unknown science of self-organization, but because the science didn't exist yet as a recognized field, their work ended up being filed on more familiar shelves. From a certain angle, those taxonomies made sense, because the leading figures of this new discipline didn't even themselves realize that they were struggling to understand the laws of emergence. They were wrestling with local issues, in clearly defined fields: how ant colonies learn to forage and built nests; why industrial neighborhoods form along class lines; how our minds learn to recognize faces. You can answer all of these questions without resorting to the sciences of complexity and self-organization, but those answers all share a common pattern, as clear as the whorls of a fingerprint. But to see it as a pattern you needed to encounter it in several contexts. Only when the pattern was detected did people begin to think about studying self-organizing systems on their own merits. Keller and Segel saw it in the slime mold assemblages; Jane Jacobs saw it in the formation of city neighborhoods; Marvin Minsky in the distributed networks of the human brain.

What features do all these systems share? In the simplest terms, they solve problems by drawing on masses of relatively stupid elements, rather than a single, intelligent "executive branch." They are bottom-up systems, not top-down. They get their smarts from below. In a more technical language, they are complex adaptive systems that display emergent behavior. In these systems, agents residing on one scale start producing behavior that lies one scale above them: ants create colonies; urbanites create neighborhoods; simple pattern-recognition software learns how to recommend new books. The movement from low-level rules to higher-level sophistication is what we call emergence.

Imagine a billiard table populated by semi-intelligent, motorized billiard balls that have been programmed to explore the space of the table and alter their movement patterns based on specific interactions with other balls. For the most part, the table is in permanent motion, with balls colliding constantly, switching directions and speed every second. Because they are motorized, they never slow down unless their rules instruct them to, and their programming enables them to take unexpected turns when they encounter other balls. Such a system would define the most elemental form of *complex* behavior: a system with multiple agents dynamically interacting in multiple ways, following local rules and oblivious to any higher-level instructions. But it wouldn't truly be considered *emergent* until those local interactions resulted in some kind of discernible macrobehavior. Say the local rules of behavior followed by the balls ended up dividing the table into two clusters of even-numbered and odd-numbered balls. That would mark the beginnings of emergence, a higher-level pattern arising out of parallel complex interactions between local agents. The balls aren't programmed explicitly to cluster in two groups; they're programmed to follow much more random rules: swerve left when they collide with a solid-colored; accelerate after contact with the three ball; stop dead in their tracks when they hit the eight ball; and so on. Yet out of those low-level routines, a coherent shape emerges.

Does that make our mechanized billiard table *adaptive*? Not really, because a table divided between two clusters of balls is not terribly useful, either to the billiard balls themselves or to anyone else in the pool hall. But, like the proverbial *Hamlet*-writing monkeys, if we had an infinite number of tables in our pool hall, each following a different set of rules, one of those tables might randomly hit upon a rule set that would arrange all the balls in a perfect triangle, leaving the cue ball across the table ready for the break. That would be adaptive behavior in the larger ecosystem of the pool hall, assuming that it was in the interest of our billiards system to attract players. The system would use local rules between interacting agents to create higher-level behavior well suited to its environment.

Emergent complexity without adaptation is like the intricate crystals formed by a snowflake: it's a beautiful pattern, but it has no function. The forms of emergent behavior that we'll examine in this book show the distinctive quality of growing smarter over time, and of responding to the specific and changing needs of their environment. In that sense, most of the systems we'll look at are more *dynamic* than our adaptive billiards table: they rarely settle in on a single, frozen shape; they form patterns in time as well as space. A better example might be a table that self-organizes into a billiards-based timing device: with the cue ball bouncing off the eight ball sixty times a minute, and the remaining balls shifting from one side of the table to another every hour on the hour. That might sound like an unlikely system to emerge out of local interactions between individual balls, but your body contains numerous organic clocks built out of simple cells that function in remarkably similar ways. An infinite number of cellular or billiard-ball configurations will not produce a working clock, and only a tiny number will. So the question becomes, how do you push your emergent system toward clocklike behavior, if that's your goal? How do you make a self-organizing system more adaptive?

That question has become particularly crucial, because the history of emergence has entered a new phase in the past few years, one that should prove to be more revolutionary than the two phases before it. In the first phase, inquiring minds struggled to understand the forces of self-organization without realizing what they were up against. In the second, certain sectors of the scientific community began to see self-organization as a problem that transcended local disciplines and set out to solve that problem, partially by comparing behavior in one area to behavior in another. By watching the slime mold cells next to the ant colonies, you could see the shared behavior in ways that would have been unimaginable watching either on its own. Self-organization became an object of study in its own right, leading to the creation of celebrated research centers such as the Santa Fe Institute, which devoted itself to the study of complexity in all its diverse forms.

But in the third phase—the one that began sometime in the past decade, the one that lies at the very heart of this book—we stopped analyzing emergence and started creating it. We began building self-organizing systems into our software applications, our video games, our art, our music. We built emergent systems to recommend new books, recognize our voices, or find mates. For as long as complex organisms have been alive, they have lived under the laws of self-organization, but in recent years our day-to-day life has become overrun with *artificial* emergence: systems built with a conscious understanding of what emergence is, systems designed to exploit those laws the same way our nuclear reactors exploit the laws of atomic physics. Up to now, the philosophers of emergence have struggled to interpret the world. But they are now starting to change it.

What follows is a tour of fields that aren't usually gathered between the same book jacket covers. We'll look at computer games that simulate living ecologies; the guild system of twelfth-century Florence; the initial cell divisions that mark the very beginning of life; and software that lets you see the patterns of your own brain. What unites these different phenomena is a recurring pattern and shape: a network of self-organization, of disparate agents that unwittingly create a higher-level order. At each scale, you can see the imprint of those slime mold cells converging; at each scale, the laws of emergence hold true.

This book roughly follows the chronology of the three historical phases. The first section introduces one of the emergent world's crowning achievements—the colony behavior of social insects such as ants and termites—and then goes back to trace part of the history of the decentralized mind-set, from Engels on the streets of Manchester to the new forms of emergent software being developed today. The second section is an overview of emergence as we currently understand it; each of the four chapters in the section explores one of the field's core principles: neighbor interaction, pattern recognition, feedback, and indirect control. The final section looks to the future of artificial emergence and speculates on what will happen when our media experiences and political movements are largely shaped by bottom-up forces, and not top-down ones.

Certain shapes and patterns hover over different moments in time, haunting and inspiring the individuals living through those periods. The epic clash and subsequent resolution of the dialectic animated the first half of the nineteenth century; the Darwinian and social reform movements scattered web imagery through the second half of the century. The first few decades of the twentieth century found their ultimate expression in the exuberant anarchy of the explosion, while later decades lost themselves in the faceless regimen of the grid. You can see the last ten years or so as a return to those Victorian webs, though I suspect the image that has been burned into our retinas over the past decade is more prosaic: windows piled atop one another on a screen, or perhaps a mouse clicking on an icon.

These shapes are shorthand for a moment in time, a way of evoking an era and its peculiar obsessions. For individuals living within these periods, the shapes are cognitive building blocks, tools for thought: Charles Darwin and George Eliot used the web as a way of understanding biological evolution and social struggles; a half century later, the futurists embraced the explosions of machine-gun fire, while Picasso used them to re-create the horrors of war in *Guernica.* The shapes are a way of interpreting the world, and while no shape completely represents its epoch, they are an undeniable component of the history of thinking.

When I imagine the shape that will hover above the first half of the twenty-first century, what comes to mind is not the coiled embrace of the genome, or the etched latticework of the silicon chip. It is instead the pulsing red and green pixels of Mitch Resnick's slime mold simulation, moving erratically across the screen at first, then slowly coalescing into larger forms. The shape of those clusters—with their lifelike irregularity, and their absent pacemakers—is the shape that will define the coming decades. I see them on the screen, growing and dividing, and I think: That way lies the future.

PART ONE

African anthill *(Courtesy of Corbis)*

Rise up, thou monstrous ant-hill on the plain
Of a too busy world! Before me flow,
Thou endless stream of men and moving things!
Thy every-day appearance, as it strikes—
With wonder heightened, or sublimed by awe—
On strangers, of all ages; the quick dance
Of colours, lights, and forms; the deafening din;
The comers and the goers face to face,
Face after face . . .

—Wordsworth,
"Residence in London"

Cities have no central planning commissions that solve the problem of purchasing and distributing supplies. . . . How do these cities avoid devastating swings between shortage and glut, year after year, decade after decade? The mystery deepens when we observe the kaleidoscopic nature of large cities. Buyers, sellers, administrations, streets, bridges, and buildings are always changing, so that a city's coherence is somehow imposed on a perpetual flux of people and structures. Like the standing wave in front of a rock in a fast-moving stream, a city is a pattern in time.

—John Holland

1
The Myth of the Ant Queen

It's early fall in Palo Alto, and Deborah Gordon and I are sitting in her office in Stanford's Gilbert Biological Sciences building, where she spends three-quarters of the year studying behavioral ecology. The other quarter is spent doing fieldwork with the native harvester ants of the American Southwest, and when we meet, her face still retains the hint of a tan from her last excursion to the Arizona desert.

I've come here to learn more about the collective intelligence of ant colonies. Gordon, dressed neatly in a white shirt, cheerfully entertains a few borderline-philosophical questions on group behavior and complex systems, but I can tell she's hankering to start with a hands-on display. After a few minutes of casual rumination, she bolts up out of her chair. "Why don't we start with me showing you the ants that we have here," she says. "And then we can talk about what it all means."

She ushers me into a sepulchral room across the hallway, where three long tables are lined up side by side. The initial impression is that of an underpopulated and sterilized pool hall, until I get close enough to one of the tables to make out the miniature civilization that lives within each of them. Closer to a Habitrail than your traditional idea of an ant farm, Gordon's contraptions house an intricate network of plastic tubes connecting a dozen or so plastic boxes, each lined with moist plaster and coated with a thin layer of dirt.

"We cover the nests with red plastic because some species of ants don't see red light," Gordon explains. "That seems to be true of this species too." For a second, I'm not sure what she means by "this species"—and then my eyes adjust to the scene, and I realize with a start that the dirt coating the plastic boxes is, in fact, thousands of harvester ants, crammed so tightly into their quarters that I had originally mistaken them for an undifferentiated mass. A second later, I can see that the whole simulated colony is wonderfully alive, the clusters of ants pulsing steadily with movement. The tubing and cramped conditions and surging crowds bring one thought immediately to mind: the New York subway system, rush hour.

At the heart of Gordon's work is a mystery about how ant colonies develop, a mystery that has implications extending far beyond the parched earth of the Arizona desert to our cities, our brains, our immune systems—and increasingly, our technology. Gordon's work focuses on the connection between the microbehavior of individual ants and the overall behavior of the colonies themselves, and part of that research involves tracking the life cycles of individual colonies, following them year after year as they scour the desert floor for food, competing with other colonies for territory, and—once a year—mating with them. She is a student, in other words, of a particular kind of emergent, self-organizing system.

Dig up a colony of native harvester ants and you'll almost invariably find that the queen is missing. To track down the colony's matriarch, you need to examine the bottom of the hole you've just dug to excavate the colony: you'll find a narrow, almost invisible passageway that leads another two feet underground, to a tiny vestibule burrowed out of the earth. There you will find the queen. She will have been secreted there by a handful of ladies-in-waiting at the first sign of disturbance. That passageway, in other words, is an emergency escape hatch, not unlike a fallout shelter buried deep below the West Wing.

But despite the Secret Service–like behavior, and the regal nomenclature, there's nothing hierarchical about the way an ant colony does its thinking. "Although *queen* is a term that reminds us of human political systems," Gordon explains, "the queen is not an authority figure. She lays eggs and is fed and cared for by the workers. She does not decide which worker does what. In a harvester ant colony, many feet of intricate tunnels and chambers and thousands of ants separate the queen, surrounded by interior workers, from the ants working outside the nest and using only the chambers near the surface. It would be physically impossible for the queen to direct every worker's decision about which task to perform and when." The harvester ants that carry the queen off to her escape hatch do so not because they've been ordered to by their leader; they do it because the queen ant is responsible for giving birth to all the members of the colony, and so it's in the colony's best interest—and the colony's gene pool—to keep the queen safe. Their genes instruct them to protect their mother, the same way their genes instruct them to forage for food. In other words, the matriarch doesn't train her servants to protect her, evolution does.

Popular culture trades in Stalinist ant stereotypes—witness the authoritarian colony regime in the animated film *Antz*—but in fact, colonies are the exact opposite of command economies. While they are capable of remarkably coordinated feats of task allocation, there are no Five-Year Plans in the ant kingdom. The colonies that

Gordon studies display some of nature's most mesmerizing decentralized behavior: intelligence and personality and learning that emerges from the bottom up.

I'm still gazing into the latticework of plastic tubing when Gordon directs my attention to the two expansive white boards attached to the main colony space, one stacked on top of the other and connected by a ramp. (Imagine a two-story parking garage built next to a subway stop.) A handful of ants meander across each plank, some porting crumblike objects on their back, others apparently just out for a stroll. If this is the Central Park of Gordon's ant metropolis, I think, it must be a workday.

Gordon gestures to the near corner of the top board, four inches from the ramp to the lower level, where a pile of strangely textured dust—littered with tiny shells and husks—presses neatly against the wall. "That's the midden," she says. "It's the town garbage dump." She points to three ants marching up the ramp, each barely visible beneath a comically oversize shell. "These ants are on midden duty: they take the trash that's left over from the food they've collected—in this case, the seeds from stalk grass—and deposit it in the midden pile."

Gordon takes two quick steps down to the other side of the table, at the far end away from the ramp. She points to what looks like another pile of dust. "And this is the cemetery." I look again, startled. She's right: hundreds of ant carcasses are piled atop one another, all carefully wedged against the table's corner. It looks brutal, and yet also strangely methodical.

I know enough about colony behavior to nod in amazement. "So they've somehow collectively decided to utilize these two areas as trash heap and cemetery," I say. No individual ant defined those areas, no central planner zoned one area for trash, the other for the dead. "It just sort of happened, right?"

Gordon smiles, and it's clear that I've missed something. "It's better than that," she says. "Look at what actually happened here: they've built the cemetery at exactly the point that's furthest away from the colony. And the midden is even more interesting: they've put it at precisely the point that maximizes its distance from both the colony *and* the cemetery. It's like there's a rule they're following: put the dead ants as far away as possible, and put the midden as far away as possible without putting it near the dead ants."

I have to take a few seconds to do the geometry myself, and sure enough, the ants have got it right. I find myself laughing out loud at the thought: it's as though they've solved one of those spatial math tests that appear on standardized tests, conjuring up a solution that's perfectly tailored to their environment, a solution that might easily stump an eight-year-old human. The question is, who's doing the conjuring?

It's a question with a long and august history, one that is scarcely limited to the collective behavior of ant colonies. We know the answer now because we have developed powerful tools for thinking about—and modeling—the emergent intelligence of self-organizing systems, but that answer was not always so clear. We know now that systems like ant colonies don't have real leaders, that the very idea of an ant "queen" is misleading. But the desire to find pacemakers in such systems has always been powerful—in both the group behavior of the social insects, and in the collective human behavior that creates a living city.

Records exist of a Roman fort dating back to A.D. 76 situated at the confluence of the Medlock and Irwell Rivers, on the northwestern edge of modern England, about 150 miles from London. Settlements persisted there for three centuries, before dying out with the rest of the empire around A.D. 400. Historians believe that the site was unoccupied for half a millennium, until a town called Manchester began to take shape there, the name derived from the Roman settlement Mamucium—Latin for "place of the breastlike hill."

Manchester subsisted through most of the millennium as a nondescript northern-England borough: granted a charter in 1301, the town established a college in the early 1400s, but remained secondary to the neighboring town of Salford for hundreds of years. In the 1600s, the Manchester region became a node for the wool trade, its merchants shipping goods to the Continent via the great ports of London. It was impossible to see it at the time, but Manchester—and indeed the entire Lancashire region—had planted itself at the very center of a technological and commercial revolution that would irrevocably alter the future of the planet. Manchester lay at the confluence of several world-historical rivers: the nascent industrial technologies of steam-powered looms; the banking system of commercial London; the global markets and labor pools of the British Empire. The story of that convergence has been told many times, and the debate over its consequences continues to this day. But beyond the epic effects that it had on the global economy, the industrial takeoff that occurred in Manchester between 1700 and 1850 also created a new kind of city, one that literally exploded into existence.

The statistics on population growth alone capture the force of that explosion: a 1773 estimate had 24,000 people living in Manchester; the first official census in 1801 found 70,000. By the midpoint of the century, there were more than 250,000 people in the city proper—a tenfold increase in only seventy-five years. That growth rate was as unprecedented and as violent as the steam engines themselves. In a real sense, the city grew too fast for the authorities to keep up with it. For five hundred years, Manchester had technically been considered a "manor," which meant, in the eyes of the law, it was run like a feudal estate, with no local government to speak of—no city planners, police, or public health authorities. Manchester didn't even send representatives to Parliament until 1832, and it wasn't incorporated for another six years. By the early 1840s, the newly formed borough council finally began to institute public health reforms and urban planning, but the British government didn't officially recognize Manchester as a city until 1853. This constitutes one of the great ironies of the industrial revolution, and it captures just how dramatic the rate of change really was: the city that most defined the future of urban life for the first half of the nineteenth century didn't legally become a city until the great explosion had run its course.

The result of that discontinuity was arguably the least planned and most chaotic city in the six-thousand-year history of urban settlements. Noisy, polluted, massively overcrowded, Manchester attracted a steady stream of intellectuals and public figures in the 1830s, traveling north to the industrial magnet in search of the modern world's future. One by one, they returned with stories of abject squalor and sensory overload, their words straining to convey the immensity and uniqueness of the experience. "What I have seen has disgusted and astonished me beyond all measure," Dickens wrote after a visit in the fall of 1838. "I mean to strike the heaviest blow in my power for these unfortunate creatures." Appointed to command the northern districts in the late 1830s, Major General Charles James Napier wrote: "Manchester is the chimney of the world. Rich rascals, poor rogues, drunken ragamuffins and prostitutes form the moral. . . . What a place! The entrance to hell, realized." De Toqueville visited Lancashire in 1835 and described the landscape in language that would be echoed throughout the next two centuries: "From this foul drain the greatest stream of human industry flows out to fertilize the whole world. From this filthy sewer pure gold flows. Here humanity attains its most complete development and its most brutish; here civilization works its miracles, and civilized man is turned back almost into a savage."

But Manchester's most celebrated and influential documentarian was a young man named Friedrich Engels, who arrived in 1842 to help oversee the family cotton plant there, and to witness firsthand the engines of history bringing the working class closer to self-awareness. While Engels was very much on the payroll of his father's firm, Ermen and Engels, by the time he arrived in Manchester he was also under the sway of the radical politics associated with the Young Hegelian school. He had befriended Karl Marx a few years before and had been encouraged to visit Manchester by the socialist Moses Hess, whom he'd met in early 1842. His three years in England were thus a kind of scouting mission for the revolution, financed by the capitalist class. The book that Engels eventually wrote, *The Condition of the Working Class in England,* remains to this day one of the classic tracts of urban history and stands as the definitive account of nineteenth-century Manchester life in all its tumult and dynamism. Dickens, Carlyle, and Disraeli had all attempted to capture Manchester in its epic wildness, but their efforts were outpaced by a twenty-four-year-old from Prussia.

But *The Condition* is not, as might be expected, purely a document of Manchester's industrial chaos, a story of all that is solid melting into air, to borrow a phrase Engels's comrade would write several years later. In the midst of the city's insanity, Engels's eye is drawn to a strange kind of order, in a wonderful passage where he leads the reader on a walking tour of the industrial capital, a tour that reveals a kind of politics built into the very topography of the city's streets. It captures Engels's acute powers of observation, but I quote from it at length because it captures something else as well—how difficult it is to think in models of self-organization, to imagine a world without pacemakers.

> The town itself is peculiarly built, so that someone can live in it for years and travel into it and out of it daily without ever com-

ing into contact with a working-class quarter or even with workers—so long, that is to say, as one confines himself to his business affairs or to strolling about for pleasure. This comes about mainly in the circumstances that through an unconscious, tacit agreement as much as through conscious, explicit intention, the working-class districts are most sharply separated from the parts of the city reserved for the middle class. . . .

I know perfectly well that this deceitful manner of building is more or less common to all big cities. I know as well that shopkeepers must in the nature of the business take premises on the main thoroughfares. I know in such streets there are more good houses than bad ones, and that the value of land is higher in their immediate vicinity than in neighborhoods that lie at a distance from them. But at the same time I have never come across so systematic a seclusion of the working class from the main streets as in Manchester. I have never elsewhere seen a concealment of such fine sensibility of everything that might offend the eyes and nerves of the middle classes. And yet it is precisely Manchester that has been built less according to a plan and less within the limitations of official regulations—and indeed more through accident—than any other town. Still . . . I cannot help feeling that the liberal industrialists, the Manchester "bigwigs," are not so altogether innocent of this bashful style of building.

You can almost hear the contradictions thundering against each other in this passage, like the "dark satanic mills" of Manchester itself. The city has built a *cordon sanitaire* to separate the industrialists from the squalor they have unleashed on the world, concealing the demoralization of Manchester's working-class districts—and yet that disappearing act comes into the world without "conscious, explicit intention." The city seems artfully planned to hide its atrocities, and yet it "has been built less according to a plan" than any city in history. As Steven Marcus puts it, in his history of the young Engels's sojourn in Manchester, "The point to be taken is that this astonishing and outrageous arrangement cannot fully be understood as the result of a plot, or even a deliberate design, although those in whose interests it works also control it. It is indeed too huge and too complex a state of organized affairs ever to have been *thought up* in advance, to have preexisted as an idea."

Those broad, glittering avenues, in other words, suggest a Potemkin village without a Potemkin. That mix of order and anarchy is what we now call emergent behavior. Urban critics since Lewis Mumford and Jane Jacobs have known that cities have lives of their own, with neighborhoods clustering into place without any Robert Moses figure dictating the plan from above. But that understanding has entered the intellectual mainstream only in recent years—when Engels paced those Manchester streets in the 1840s, he was left groping blindly, trying to find a culprit for the city's fiendish organization, even as he acknowledged that the city was notoriously unplanned. Like most intellectual histories, the development of that new understanding—the sciences of complexity and self-organization—is a complicated, multithreaded tale, with many agents interacting over its duration. It is probably better to think of it as less a linear narrative and more an interconnected web, growing increasingly dense over the century and a half that separates us from Engels's first visit to Manchester.

Complexity is a word that has frequently appeared in critical accounts of metropolitan space, but there are really two kinds of complexity fundamental to the city, two experiences with very different implications for the individuals trying to make sense of them. There is, first, the more conventional sense of complexity as sensory overload, the city stretching the human nervous system to its very extremes, and in the process teaching it a new series of reflexes—and leading the way for a complementary series of aesthetic values, which develop out like a scab around the original wound. The German cultural critic Walter Benjamin writes in his unfinished masterpiece, *The Arcades Project*:

> Perhaps the daily sight of a moving crowd once presented the eye with a spectacle to which it first had to adapt. . . . [T]hen the assumption is not impossible that, having mastered this task, the eye welcomed opportunities to confirm its possession of its new ability. The method of impressionist painting, whereby the picture is assembled through a riot of flecks of color, would then be a reflection of experience with which the eye of a big-city dweller has become familiar.

There's a long tributary of nineteenth- and twentieth-century urban writing that leads into this passage, from the London chapters of Wordsworth's *Prelude* to the ambulatory musings of Joyce's *Dubliners*: the noise and the senselessness somehow transformed into an aesthetic experience. The crowd is something you throw yourself into, for the pure poetry of it all. But complexity is not solely a matter of sensory overload. There is also the sense of complexity as a self-organizing system—more Santa Fe Institute than Frankfurt School. This sort of complexity lives up one level: it describes the system of the city itself, and not its experiential reception by the city dweller. The city is complex because it overwhelms, yes, but also because it has a coherent personality, a personality that self-organizes out of millions of individual decisions, a global order built out of local interactions. This is the "systematic" complexity that Engels glimpsed on the boulevards of Manchester: not the overload and anarchy he documented elsewhere, but instead a strange kind of order, a pattern in the streets that furthered the political values of

Manchester's elite without being deliberately planned by them. We know now from computer models and sociological studies—as well as from the studies of comparable systems generated by the social insects, such as Gordon's harvester ants—that larger patterns can emerge out of uncoordinated local actions. But for Engels and his contemporaries, those unplanned urban shapes must have seemed like a haunting. The city appeared to have a life of its own.

A hundred and fifty years later, the same techniques translated into the language of software—as in Mitch Resnick's slime mold simulation—trigger a similar reaction: the eerie sense of something lifelike, something organic forming on the screen. Even those with sophisticated knowledge about self-organizing systems still find these shapes unnerving—in their mix of stability and change, in their capacity for open-ended learning. The impulse to build centralized models to explain that behavior remains almost as strong as it did in Engels's day. When we see repeated shapes and structure emerging out of apparent chaos, we can't help looking for pacemakers.

Understood in the most abstract sense, what Engels observed are *patterns* in the urban landscape, visible because they have a repeated structure that distinguishes them from the pure noise you might naturally associate with an unplanned city. They are patterns of human movement and decision-making that have been etched into the texture of city blocks, patterns that are then fed back to the Manchester residents themselves, altering their subsequent decisions. (In that sense, they are the very opposite of the traditional sense of urban complexity—they are signals emerging where you would otherwise expect only noise.) A city is a kind of pattern-amplifying machine: its neighborhoods are a way of measuring and expressing the repeated behavior of larger collectivities—capturing information about group behavior, and sharing that information with the group. Because those patterns are fed back to the community, small shifts in behavior can quickly escalate into larger movements: upscale shops dominate the main boulevards, while the working class remains clustered invisibly in the alleys and side streets; the artists live on the Left Bank, the investment bankers in the Eighth Arrondissement. You don't need regulations and city planners deliberately creating these structures. All you need are thousands of individuals and a few simple rules of interaction. The bright shop windows attract more bright shop windows and drive the impoverished toward the hidden core. There's no need for a Baron Haussmann in this world, just a few repeating patterns of movement, amplified into larger shapes that last for lifetimes: clusters, slums, neighborhoods.

Not all patterns are visible to every city dweller, though. The history of urbanism is also the story of more muted signs, built by the collective behavior of smaller groups and rarely detected by outsiders. Manchester harbors several such secret clusters, persisting over the course of many generations, like a "standing wave in front of a rock in a fast-moving stream." One of them lies just north of Victoria University, at a point where Oxford Road becomes Oxford Street. There are reports dating back to the mid-nineteenth century of men cruising other men on these blocks, looking for casual sex, more lasting relationships, or even just the camaraderie of shared identity at a time when that identity dared not speak its name. Some historians speculate that Wittgenstein visited these streets during his sojourn in Manchester in 1908. Nearly a hundred years later, the area has christened itself the Gay Village and actively promotes its coffee bars and boutiques as a must-see Manchester tourist destination, like Manhattan's Christopher Street and San Francisco's Castro. The pattern is now broadcast to a wider audience, but it has not lost its shape.

But even at a lower amplitude, that signal was still loud enough to attract the attention of another of Manchester's illustrious immigrants: the British polymath Alan Turing. As part of his heroic contribution to the war effort, Turing had been a student of mathematical patterns, designing the equations and the machines that cracked the "unbreakable" German code of the Enigma device. After a frustrating three-year stint at the National Physical Laboratory in London, Turing moved to Manchester in 1948 to help run the university's embryonic computing lab. It was in Manchester that Turing began to think about the problem of biological development in mathematical terms, leading the way to the "Morphogenesis" paper, published in 1952, that Evelyn Fox Keller would rediscover more than a decade later. Turing's war research had focused on detecting patterns lurking within the apparent chaos of code, but in his Manchester years, his mind gravitated toward a mirror image of the original code-breaking problem: how complex patterns could come into being by following simple rules. How does a seed know how to build a flower?

Turing's paper on morphogenesis—literally, "the beginning of shape"—turned out to be one of his seminal works, ranking up their with his more publicized papers and speculations: his work on Gödel's undecidability problem, the Turing Machine, the Turing Test—not to mention his contributions to the physical design of the modern digital computer. But the morphogenesis paper was only the beginning of a shape—a brilliant mind sensing the outlines of a new problem, but not fully grasping all its intricacies. If Turing had been granted another few decades to explore the powers of self-assembly—not to mention access to the number-crunching horsepower of non-vacuum-tube computers—it's not hard to imagine his mind greatly enhancing our subsequent understanding of emergent behavior. But the work on morphogenesis was tragically cut short by his death in 1954.

Alan Turing was most likely a casualty of the brutally homophobic laws of postwar Britain, but his death also intersected with those discreet patterns of life on Manchester's sidewalks. Turing had known about that stretch of Oxford Road since his arrival in Manchester; on occasion, he would drift down to the neighborhood, meeting other gay men—inviting some of them back to his flat for conversation, and presumably some sort of physical contact. In January of 1952, Turing met a young man named Arnold Murray on those streets, and the two embarked on a brief relationship that quickly turned sour. Murray—or a friend of Murray's—broke into Turing's house and stole a few items. Turing reported the theft to the police and, with his typical forthrightness, made no effort to conceal the affair with Murray when the police visited his flat. Homosexuality was a criminal offense according to British law, punishable by up to two years' imprisonment, and so the police promptly charged both Turing and Murray with "gross indecency."

On February 29, 1952, while the Manchester authorities were preparing their case against him, Turing finished the revisions to his morphogenesis paper, and he argued over its merits with Ilya Prigogine, the visiting Belgian chemist whose work on nonequilibrium thermodynamics would later win him a Nobel prize. In one day, Turing had completed the text that would help engender the discipline of biomathematics and inspire Keller and Segel's slime mold discoveries fifteen years later, and he had enjoyed a spirited exchange with the man who would eventually achieve world fame for his research into self-organizing systems. On that winter day in 1952, there was no mind on the face of the earth better prepared to wrestle with the mysteries of emergence than Alan Turing's. But the world outside that mind was conspiring to destroy it. That very morning, a local paper broke the story that the war-hero savant had been caught in an illicit affair with a nineteen-year-old boy.

Within a few months Turing had been convicted of the crime and placed on a humiliating estrogen treatment to "cure" him of his homosexuality. Hounded by the authorities and denied security clearance for the top-secret British computing projects he had been contributing to, Turing died two years later, an apparent suicide.

Turing's career had already collided several times with the developing web of emergence before those fateful years in Manchester. In the early forties, during the height of the war effort, he had spent several months at the legendary Bell Laboratories on Manhattan's West Street, working on a number of encryption schemes, including an effort to transmit heavily encoded waveforms that could be decoded as human speech with the use of a special key. Early in his visit to Bell Labs, Turing hit upon the idea of using another Bell invention, the Vocoder—later used by rock musicians such as Peter Frampton to combine the sounds of a guitar and the human voice—as a way of encrypting speech. (By early 1943, Turing's ideas had enabled the first secure voice transmission to cross the Atlantic, unintelligible to German eavesdroppers.) Bell Labs was the home base for another genius, Claude Shannon, who would go on to found the influential discipline of information theory, and whose work had explored the boundaries between noise and information. Shannon had been particularly intrigued by the potential for machines to detect and amplify patterns of information in noisy communication channels—a line of inquiry that promised obvious value to a telephone company, but could also save thousands of lives in a war effort that relied so heavily on the sending and breaking of codes. Shannon and Turing immediately recognized that they had been working along parallel tracks: they were both code-breakers by profession at that point, and in their attempts to build automated machines that could recognize patterns in audio signals or numerical sequences, they had both glimpsed a future populated by even more intelligence machines. Shannon and Turing passed many an extended lunchtime at the Bell Labs, trading ideas on an "electronic brain" that might be capable of humanlike feats of pattern recognition.

Turing had imagined his thinking machine primarily in terms of its logical possibilities, its ability to execute an infinite variety of computational routines. But Shannon pushed him to think of the machine as something closer to an actual human brain, capable of recognizing more nuanced patterns. One day over lunch at the lab, Turing exclaimed playfully to his colleagues, "Shannon wants to feed not just data to a brain, but *cultural* things! He wants to play music to it!" Musical notes were patterns too, Shannon recognized, and if you could train an electronic brain to understand and respond to logical patterns of zeros and ones, then perhaps sometime in the future we could train our machines to appreciate the equivalent patterns of minor chord progressions and arpeggios. The idea seemed fanciful at the time—it was hard enough getting a machine to perform long division, much less savor Beethoven's Ninth. But the pattern recognition that Turing and Shannon envisioned for digital computers has, in recent years, become a central part of our cultural life, with machines both generating music for our entertainment and recommending new artists for us to enjoy. The connection between musical patterns and our neurological wiring would play a central role in one of the founding texts of modern artificial intelligence, Douglas Hofstadter's *Gödel, Escher, Bach.* Our computers still haven't developed a genuine ear for music, but if they ever do, their skill will date back to those lunchtime conversations between Shannon and Turing at Bell Labs. And that learning too will be a kind of emergence, a higher-level order forming out of relatively simple component parts.

Five years after his interactions with Turing, Shannon published a long essay in the *Bell System Technical Journal* that was quickly repackaged as a book called *The Mathematical Theory of Communication.* Dense with equations and arcane chapter titles such as "Discrete Noiseless Systems," the book managed to become something of a cult classic, and the discipline it spawned—information theory—had a profound impact on scientific and technological research that followed, on both a theoretical and practical level. *The Mathematical Theory of Communication* contained an elegant, layman's introduction to Shannon's theory, penned by the esteemed scientist Warren Weaver, who had early on grasped the significance of Shannon's work. Weaver had played a leading role in the Natural Sciences division of the Rockefeller Foundation since 1932, and when he retired in the late fifties, he composed a long report for the foundation, looking back at the scientific progress that had been achieved over the preceding quarter century. The occasion suggested a reflective look backward, but the document that Weaver produced (based loosely on a paper he had written for *American Scientist)* was far more prescient, more forward-looking. In many respects, it deserves to be thought of as the founding text of complexity theory—the point at which the study of complex systems began to think of itself as a unified field. Drawing upon research in molecular biology, genetics, physics, computer science, and Shannon's information theory, Weaver divided the last few centuries of scientific inquiry into three broad camps. First, the study of simple systems: two or three variable problems, such as the rotation of planets, or the connection between an electric current and its voltage and resistance. Second, problems of "disorganized complexity": problems characterized by millions or billions of variables that can only be approached by the methods of statistical mechanics and probability theory. These tools helped explain not only the behavior of molecules in a gas, or the patterns of heredity in a gene pool, but also helped life insurance companies turn a profit despite their limited knowledge about any individual human's future health. Thanks to Claude Shannon's work, the statistical approach also helped phone companies deliver more reliable and intelligible long-distance service.

But there was a third phase to this progression, and we were only beginning to understand. "This statistical method of dealing with disorganized complexity, so powerful an advance over the earlier two-variable methods, leaves a great field untouched," Weaver wrote. There was a middle region between two-variable equations and problems that involved billions of variables. Conventionally, this region involved a "moderate" number of variables, but the size of the system was in fact a secondary characteristic:

> Much more important than the mere number of variables is the fact that these variables are all interrelated. . . . These problems, as contrasted with the disorganized situations with which statistics can cope, *show the essential feature of organization.* We will therefore refer to this group of problems as those of *organized complexity.*

Think of these three categories of problems in terms of our billiards table analogy from the introduction. A two- or three-variable problem would be an ordinary billiards table, with balls bouncing off one another following simple rules: their velocities, the friction of the table. That would be an example of a "simple system"—and indeed, billiard balls are often used to illustrate basic laws of physics in high school textbooks. A system of disorganized complexity would be that same table enlarged to include a million balls, colliding with one another millions of times a second. Making predictions about the behavior of any individual ball in that mix would be difficult, but you could make some accurate predictions about the overall behavior of the table. Assuming there's enough energy in the system at the outset, the balls will spread to fill the entire table, like gas molecules in a container. It's complex because there are many interacting agents, but it's disorganized because they don't create any higher-level behavior other than broad statistical trends. Organized complexity, on the other hand, is like our motorized billiards table, where the balls follow specific rules and through their various interactions create a distinct macrobehavior, arranging themselves in a specific shape, or forming a specific pattern over time. That sort of behavior, for Weaver, suggested a problem of organized complexity, a problem that suddenly seemed omnipresent in nature once you started to look for it:

> What makes an evening primrose open when it does? Why does salt water fail to satisfy thirst? . . . What is the description of aging in biochemical terms? . . . What is a gene, and how does the original genetic constitution of a living organism express itself in the developed characteristics of the adult?
>
> All these are certainly complex problems. But they are not problems of disorganized complexity, to which statistical methods hold the key. They are all problems which involve dealing simultaneously with a sizable number of factors which are interrelated into an organic whole.

Tackling such problems required a new approach: "The great central concerns of the biologist . . . are now being approached not only from *above,* with the broad view of the natural philosopher who scans the whole living world, but also from *underneath,* by the quantitative analyst who measures the underlying facts." This was a genuine shift in the paradigm of research, to use Thomas Kuhn's language—a revolution not so much in the interpretations that science built in its attempt to explain the world, but rather in the types of questions it asked. The paradigm shift was more than just a new mind-set, Weaver recognized; it was also a by-product of new tools that were appearing on the horizon. To solve the problems of organized complexity, you needed a machine capable of churning through thousands, if not millions, of calculations per second—a rate that would have been unimaginable for individual brains running the numbers with the limited calculating machines of the past few centuries. Because of his connection to the Bell Labs group, Weaver had seen early on the promise of digital computing, and he knew that the mysteries of organized complexity would be much easier to tackle once you could model the behavior in close-to-real time. For millennia, humans had used their skills at observation and classification to document the subtle anatomy of flowers, but for the first time they were perched on the brink of answering a more fundamental question, a question that had more to do with patterns developing over time than with static structure: Why does an evening primrose open when it does? And how does a simple seed know how to make a primrose in the first place?

Alan Turing had played an essential role in creating both the hardware and the software that powered this first digital revolution, and his work on morphogenesis had been one of the first systematic attempts to imagine development as a problem of organized complexity. It is one of the great tragedies of this story that Turing didn't live to see—much less participate in—the extraordinary intellectual flowering that took place when those two paths intersected.

Ironically, Warren Weaver's call to action generated the first major breakthrough in a work that had nothing to do with digital computers—a work that belonged to a field not usually considered part of the hard sciences. In the years after the war, urban planners and government officials had been tackling the problem of inner-city slums with a decidedly top-down approach: razing entire neighborhoods and building bleak high-rise housing projects, ringed by soon-to-be-derelict gardens and playgrounds. The projects effectively tried to deal with the problem of dangerous city streets by eliminating streets altogether, and while the apartments in these new high-rises usually marked an improvement in living space and infrastructure, the overall environment of the projects quickly descended into an anonymous war zone that managed to both increase the crime rate in the area and destroy the neighborhood feel that had preceded them.

In October of 1961, the New York City Planning Commission announced its findings that a large portion of the historic West Village was "characterized by blight, and suitable for clearance, replanning, reconstruction, or rehabilitation." The Village community—a lively mix of artists, writers, Puerto Rican immigrants, and working-class Italian-Americans—responded with outrage, and at the center of the protests was an impassioned urban critic named Jane Jacobs. Jacobs had just spearheaded a successful campaign to block urban-development kingpin Robert Moses's plan to build a superhighway through the heart of SoHo, and she was now turning her attention to the madness of the projects. (The proposed "rehabilitation" included Jacobs's own residence on Hudson Street.) In her valiant and ultimately triumphant bid to block the razing of the West Village, Jacobs argued that the way to improve city streets and restore the dynamic civility of urban life was not to bulldoze the problem zones, but rather to look at city streets that did work and learn from them. Sometime in the writing of what would become *The Death and Life of the Great American Cities*—published shortly after the Village showdown—Jacobs read Warren Weaver's Rockefeller Foundation essay, and she immediately recognized her own agenda in his call for exploring problems of organized complexity.

> Under the seeming disorder of the old city, wherever the old city is working successfully, is a marvelous order for maintaining the safety of the streets and the freedom of the city. It is a complex order. Its essence is intimacy of sidewalk use, bringing with it a constant succession of eyes. This order is all composed of movement and change, and although it is life, not art, we may fancifully call it the art form of the city and liken it to the dance—not to a simple-minded precision dance with everyone kicking up at the same time, twirling in unison and bowing off en masse, but to an intricate ballet in which the individual dancers and ensembles all have distinctive parts which miraculously reinforce each other and compose an orderly whole.

Jacobs gave *Death and Life*'s closing chapter the memorable title "The Kind of Problem a City Is," and she began it by quoting extensively from Weaver's essay. Understanding how a city works, Jacobs argued, demanded that you approach it as a problem from the street level up. "In parts of cities which are working well in some respects and badly in others (as is often the case), we cannot even analyze the virtues and the faults, diagnose the trouble or consider helpful changes, without going at them as problems of organized complexity," she wrote. "We may wish for easier, all-purpose analyses, and for simpler, magical, all-purpose cures, but wishing cannot change these problems into simpler matters than organized complexity, no matter how much we try to evade the realities and to handle them as something different." To understand the city's complex order, you needed to understand that ever-changing ballet; where city streets had lost their equilibrium, you couldn't simply approach the problem by fiat and bulldoze entire neighborhoods out of existence.

Jacobs's book would revolutionize the way we imagined cities. Drawing on Weaver's insights, she conveyed a vision of the city as far more than the sum of its residents—closer to a living organism, capable of adaptive change. "Vital cities have marvelous innate abilities for understanding, communicating, contriving and inventing what is required to combat their difficulties," she wrote. They get their order from below; they are learning machines, pattern recognizers—even when the patterns they respond to are unhealthy ones. A century after Engels glimpsed the systematic disappearing act of Manchester's urban poor, the self-organizing city had finally come into focus.

"Organized complexity" proved to be a constructive way of thinking about urban life, but Jacobs's book was a work of social theory, not science. Was it possible to model and explain the behavior of self-organizing systems using more rigorous methods? Could the developing technology of digital computing be usefully applied to this problem? Partially thanks to Shannon's work in the late forties, the biological sciences had made a number of significant breakthroughs in understanding pattern recognition and feedback by the time Jacobs published her masterpiece. Shortly after his appointment to the Harvard faculty in 1956, the entomologist Edward O. Wilson convincingly proved that ants communicate with one another—and coordinate overall colony behavior—by recognizing patterns in pheromone trails left by fellow ants, not unlike the cyclic AMP signals of the slime mold. At the Free University of Brussels in the fifties, Ilya Prigogine was making steady advances in his understanding of nonequilibrium thermodynamics, environments where the laws of entropy are temporarily overcome, and higher-level order may spontaneously emerge out of underlying chaos. And at MIT's Lincoln Laboratory, a twenty-five-year-old researcher named Oliver Selfridge was experimenting with a model for teaching a computer how to learn.

There is a world of difference between a computer that passively receives the information you supply and a computer that actively learns on its own. The very first generation of computers such as ENIAC had processed information fed to them by their masters, and they had been capable of performing various calculations with that data, based on the instruction sets programmed into them. This was a startling enough development at a time when "computer" meant a person with a slide rule and an eraser. But even in those early days, the digital visionaries had imagined a machine capable of more open-ended learning. Turing and Shannon had argued over the future musical tastes of the "electronic brain" during lunch hour at Bell Labs, while their colleague Norbert Wiener had written a best-selling paean to the self-regulatory powers of feedback in his 1949 manifesto *Cybernetics.*

"Mostly my participation in all of this is a matter of good luck for me," Selfridge says today, sitting in his cramped, windowless MIT office. Born in England, Selfridge enrolled at Harvard at the age of fifteen and started his doctorate three years later at MIT, where Norbert Wiener was his dissertation adviser. As a precocious twenty-one-year-old, Selfridge suggested a few corrections to a paper that his mentor had published on heart flutters, corrections that Wiener graciously acknowledged in the opening pages of *Cybernetics.* "I think I now have the honor of being one of the few living people mentioned in that book," Selfridge says, laughing.

After a sojourn working on military control projects in New Jersey, Selfridge returned to MIT in the midfifties. His return coincided with an explosion of interest in artificial intelligence (AI), a development that introduced him to a then-junior fellow at Harvard named Marvin Minsky. "My concerns in AI," Selfridge says now, "were not so much the actual processing as they were in how systems change, how they evolve—in a word, how they learn." Exploring the possibilities of machine learning brought Selfridge back to memories of his own education in England. "At school in England I had read John Milton's *Paradise Lost,*" he says, "and I'd been struck by the image of Pandemonium—it's Greek for 'all the demons.' Then after my second son, Peter, was born, I went over *Paradise Lost* again, and the shrieking of the demons awoke something in me." The pattern recognizer in Selfridge's brain had hit upon a way of teaching a computer to recognize patterns.

"We are proposing here a model of a process which we claim can adaptively improve itself to handle certain pattern-recognition problems which cannot be adequately specified in advance." These were the first words Selfridge delivered at a symposium in late 1958, held at the very same National Physical Laboratory from which Turing had escaped a decade before. Selfridge's presentation had the memorable title "Pandemonium: A Paradigm for Learning," and while it had little impact outside the nascent computer-science community, the ideas Selfridge outlined that day would eventually become part of our everyday life—each time we enter a name in our PalmPilots or use voice-recognition software to ask for information over the phone. Pandemonium, as Selfridge outlined it in his talk, was not so much a specific piece of software as it was a way of approaching a problem. The problem was an ambitious one, given the limited computational resources of the day: how to teach a computer to recognize patterns that were ill-defined or erratic, like the sound waves that comprise spoken language.

The brilliance of Selfridge's new paradigm lay in the fact that it relied on a distributed, bottom-up intelligence, and not a unified, top-down one. Rather than build a single smart program, Selfridge created a swarm of limited miniprograms, which he called demons. "The idea was, we have a bunch of these demons shrieking up the hierarchy," he explains. "Lower-level demons shrieking to higher-level demons shrieking to higher ones."

To understand what that "shrieking" means, imagine a system

with twenty-six individual demons, each trained to recognize a letter of the alphabet. The pool of demons is shown a series of words, and each demon "votes" as to whether each letter displayed represents its chosen letter. If the first letter is *a,* the *a*-recognizing demon reports that it is highly likely that it has recognized a match. Because of the similarities in shape, the *o*-recognizer might report a possible match, while the *b*-recognizer would emphatically declare that the letter wasn't intelligible to it. All the letter-recognizing demons would report to a master demon, who would tally up the votes for each letter and choose the demon that expressed the highest confidence. Then the software would move on to the next letter in the sequence, and the process would begin again. At the end of the transmission, the master demon would have a working interpretation of the text that had been transmitted, based on the assembled votes of the demon democracy.

Of course, the accuracy of that interpretation depended on the accuracy of the letter recognizers. If you were trying to teach a computer how to read, it was cheating to assume from the outset that you could find twenty-six accurate letter recognizers. Selfridge was after a larger goal: How do you teach a machine to recognize letters—or vowel sounds, minor chords, fingerprints—in the first place? The answer involved adding another layer of demons, and a feedback mechanism whereby the various demon guesses could be graded. This lower level was populated by even less sophisticated miniprograms, trained only to recognize raw physical shapes (or sounds, in the case of Morse code or spoken language). Some demons recognized parallel lines, others perpendicular ones. Some demons looked for circles, others for dots. None of these shapes were associated with any particular letter; these bottom-dwelling demons were like two-year-old children—capable of reporting on the shapes they witnessed, but not perceiving them as letters or words.

Using these minimally equipped demons, the system could be trained to recognize letters, without "knowing" anything about the alphabet in advance. The recipe was relatively simple: Present the letter *b* to the bottom-level demons, and see which ones respond, and which ones don't. In the case of the letter *b,* the vertical-line recognizers might respond, along with the circle recognizers. Those lower-level demons would report to a letter-recognizer one step higher in the chain. Based on the information gathered from its lieutenants, that recognizer would make a guess as to the letter's identity. Those guesses are then "graded" by the software. If the guess is wrong, the software learns to dissociate those particular lieutenants from the letter in question; if the guess happens to be right, it *strengthens* the connection between the lieutenants and the letter.

The results are close to random at first, but if you repeat the process a thousand times, or ten thousand, the system learns to associate specific assembles of shape-recognizers with specific letters and soon enough is capable of translating entire sentences with remarkable accuracy. The system doesn't come with any predefined conceptions about the shapes of letters—you train the system to associate letters with specific shapes in the grading phase. (This is why handwriting-recognition software can adapt to so many different types of penmanship, but *can't* adapt to penmanship that changes day to day.) That mix of random beginnings organizing into more complicated results reminded Selfridge of another process, whose own underlying code was just then being deciphered in the form of DNA. "The scheme sketched is really a natural selection on the processing demons," Selfridge explained. "If they serve a useful function they survive and perhaps are even the source for other subdemons who are themselves judged on their merits. It is perfectly reasonable to conceive of this taking place on a broader scale . . . instead of having but one Pandemonium we might have some crowd of them, all fairly similarly constructed, and employ natural selection on the crowd of them."

The system Selfridge described—with its bottom-up learning, and its evaluating-feedback loops—belongs in the history books as the first practical description of an emergent software program. The world now swarms with millions of his demons.

Among the students at MIT in the late forties was a transplanted midwesterner named John Holland. Holland was also a pupil of Norbert Wiener's, and he spent a great deal of his undergraduate years stealing time on the early computer prototypes being built in Cambridge at that time. His unusual expertise at computer programming led IBM to hire him in the fifties to help develop their first commercial calculator, the 701. As a student of Wiener's, he was naturally inclined to experiment with ways to make the sluggish 701 machine learn in a more organic, bottom-up fashion—not unlike Selfridge's Pandemonium—and Holland and a group of like-minded colleagues actually programmed a crude simulation of neurons interacting. But IBM was in the business of selling adding machines then, and so Holland's work went largely ignored and underfunded. After a few years Holland returned to academia to get his doctorate at the University of Michigan, where the Logic of Computers Group had just been formed.

In the sixties, after graduating as the first computer science Ph.D. in the country, Holland began a line of inquiry that would dominate his work for the rest of his life. Like Turing, Holland wanted to explore the way simple rules could lead to complex behavior; like Selfridge, he wanted to create software that would be capable of open-ended learning. Holland's great breakthrough was to harness the forces of another bottom-up, open-ended system: natural selection. Building on Selfridge's Pandemonium model,

Holland took the logic of Darwinian evolution and built it into code. He called his new creation the genetic algorithm.

A traditional software program is a series of instructions that tells the computer what to do: paint the screen with red pixels, multiply a set of numbers, delete a file. Usually those instructions are encoded as a series of branching paths: do this first, and if you get result A, do one thing; if you get result B, do another thing. The art of programming lay in figuring out how to construct the most efficient sequence of instructions, the sequence that would get the most done with the shortest amount of code—and with the least likelihood of a crash. Normally that was done using the raw intellectual firepower of the programmer's mind. You thought about the problem, sketched out the best solution, fed it into the computer, evaluated its success, and then tinkered with it to make it better. But Holland imagined another approach: set up a gene pool of possible software and let successful programs *evolve* out of the soup.

Holland's system revolved around a series of neat parallels between computer programs and earth's life-forms. Each depends on a master code for its existence: the zeros and ones of computer programming, and the coiled strands of DNA lurking in all of our cells (usually called the genotype). Those two kinds of codes dictate some kind of higher-level form or behavior (the phenotype): growing red hair or multiplying two numbers together. With DNA-based organisms, natural selection works by creating a massive pool of genetic variation, then evaluating the success rate of the assorted behaviors unleashed by all those genes. Successful variations get passed down to the next generation, while unsuccessful ones disappear. Sexual reproduction ensures that the innovative combinations of genes find each other. Occasionally, random mutations appear in the gene pool, introducing complete new avenues for the system to explore. Run through enough cycles, and you have a recipe for engineering masterworks like the human eye—without a bona fide engineer in sight.

The genetic algorithm was an attempt to capture that process in silicon. Software already has a genotype and a phenotype, Holland recognized; there's the code itself, and then there's what the code actually *does.* What if you created a gene pool of different code combinations, then evaluated the success rate of the phenotypes, eliminating the least successful strands? Natural selection relies on a brilliantly simple, but somewhat tautological, criterion for evaluating success: your genes get to pass on to the next generation if you survive long enough to produce a next generation. Holland decided to make that evaluation step more precise: his programs would be admitted to the next generation if they did a better job of accomplishing a specific task—doing simple math, say, or recognizing patterns in visual images. The programmer could decide what the task was; he or she just couldn't directly instruct the software how to accomplish it. He or she would set up the parameters that defined genetic fitness, then let the software evolve on its own.

Holland developed his ideas in the sixties and seventies using mostly paper and pencil—even the more advanced technology of that era was far too slow to churn through the thousandfold generations of evolutionary time. But the massively parallel, high-speed computers introduced in the eighties—such as Danny Hillis's Connection Machine—were ideally suited for exploring the powers of the genetic algorithm. And one of the most impressive GA systems devised for the Connection Machine focused exclusively on simulating the behavior of ants.

It was a program called Tracker, designed in the mideighties by two UCLA professors, David Jefferson and Chuck Taylor. (Jefferson was in the computer science department, while Taylor was a biologist.) "I got the idea from reading Richard Dawkins's first book, *The Selfish Gene,*" Jefferson says today. "That book really transformed me.

He makes the point that in order to watch Darwinian evolution in action, all you need are objects that are capable of reproducing themselves, and reproducing themselves imperfectly, and having some sort of resource limitation so that there's competition. And nothing else matters—it's a very tiny, abstract axiom that is required to make evolution work. And so it occurred to me that programs have those properties—programs can reproduce themselves. Except that they usually reproduce themselves *exactly*. But I recognized that if there was a way to have them reproduce imperfectly, and if you had not just one program but a whole population of them, then you could simulate evolution with the software instead of organisms."

After a few small-scale experiments, Jefferson and Taylor decided to simulate the behavior of ants learning to follow a pheromone trail. "Ants were on my mind—I was looking for simple creatures, and E. O. Wilson's opus on ants had just come out," Jefferson explains. "What we were really looking for was a simple task that simple creatures perform where it wasn't obvious how to make a program do it. Somehow we came up with the idea of following a trail—and not just a clean trail, a noisy trail, a broken trail." The two scientists created a virtual grid of squares, drawing a meandering path of eighty-two squares across it. Their goal was to evolve a simple program, a virtual ant, that could navigate the length of the path in a finite amount of time, using only limited information about the path's twists and turns. At each cycle, an ant had the option of "sniffing" the square ahead of him, advancing forward one square, or turning right or left ninety degrees. Jefferson and Taylor gave their ants one hundred cycles to navigate the path; once an ant used up his hundred cycles, the software tallied up the number of squares on the trail he had successfully landed on and gave him a score. An ant that lost his way after square one would be graded 1; an ant that successfully completed the trail before the hundred cycles were up would get a perfect score, 82.

The scoring system allowed Jefferson and Taylor to create fitness criteria that determined which ants were allowed to reproduce. Tracker began by simulating sixteen thousand ants—one for each of the Connection Machine's processors—with sixteen thousand more or less random strategies for trail navigation. One ant might begin with the strategy of marching straight across the grid; another by switching back and forth between ninety-degree rotations and sniffings; another following more baroque rules. The great preponderance of these strategies would be complete disasters, but a few would allow a stumble across a larger portion of the trail. Those more successful ants would be allowed to mate and reproduce, creating a new generation of sixteen thousand ants ready to tackle the trail.

The path—dubbed the John Muir Trail after the famous environmentalist—began with a relatively straightforward section, with a handful of right-hand turns and longer straight sections, then steadily grew more complicated. Jefferson says now that he designed it that way because he was worried that early generations would be so incompetent that a more challenging path would utterly confound them. "You have to remember that we had no idea when we started this experiment whether sixteen thousand was anywhere near a large enough population to seek Darwinian evolution," he explains. "And I didn't know if it was going to take ten generations, or one hundred generations, or ten thousand generations. There was no theory to guide us quantitatively about either the size of the population in space or the length of the experiment in time."

Running through one hundred generations took about two hours; Jefferson and Taylor rigged the system to give them real-time updates on the most talented ants of each generation. Like a stock ticker, the Connection Machine would spit out an updated number at the end of each generation: if the best trail-follower of one generation managed to hit fifteen squares in a hundred cycles, the Connection Machine would report that 15 was the current record and then move on to the next generation. After a few false starts because of bugs, Jefferson and Taylor got the Tracker system to work—and the results exceeded even their most optimistic expectations.

"To our wonderment and utter joy," Jefferson recalls, "it succeeded the first time. We were sitting there watching these numbers come in: one generation would produce twenty-five, then twenty-five, and then it would be twenty-seven, and then thirty. Eventually we saw a perfect score, after only about a hundred generations. It was mind-blowing." The software had evolved an entire population of expert trail-followers, despite the fact that Jefferson and Taylor had endowed their first generation of ants with no skills whatsoever. Rather than engineer a solution to the trail-following problem, the two UCLA professors had evolved a solution; they had created a random pool of possible programs, then built a feedback mechanism that allowed more successful programs to emerge. In fact, the evolved programs were so successful that they'd developed solutions custom-tailored to their environments. When Jefferson and Taylor "dissected" one of the final champion ants to see what trail-following strategies he had developed, they discovered that the software had evolved a preference for making right-hand turns, in response to the three initial right turns that Jefferson had built into the John Muir Trail. It was like watching an organism living in water evolving gills: even in the crude, abstract grid of Tracker, the virtual ants evolved a strategy for survival that was uniquely adapted to their environment.

By any measure, Tracker was a genuine breakthrough. Finally the tools of modern computing had advanced to the point where you could simulate emergent intelligence, watch it unfold on the screen in real time, as Turing and Selfridge and Shannon had dreamed of doing years before. And it was only fitting that Jefferson and Taylor had chosen to simulate precisely the organism most celebrated for its emergent behavior: the ant. They began, of course, with the most elemental form of ant intelligence—sniffing for pheromone trails—but the possibilities suggested by the success of Tracker were endless. The tools of emergent software had been harnessed to model and understand the evolution of emergent intelligence in real-world organisms. In fact, watching those virtual ants evolve on the computer screen, learning and adapting to their environments on their own, you couldn't help wonder if the division between the real and the virtual was becoming increasingly hazy.

In Mitch Resnick's computer simulation of slime mold behavior, there are two key variables, two elements that you can alter in your interaction with the simulation. The first is the number of slime mold cells in the system; the second is the physical and temporal length of the pheromone trail left behind by each cell as it crawls across the screen. (You can have long trails that take minutes to evaporate, or short ones that disappear within seconds.) Because slime mold cells collectively decide to aggregate based on their encounters with pheromone trails, altering these two variables can have a massive impact on the simulated behavior of the system. Keep the trails short and the cells few, and the slime molds will steadfastly refuse to come together. The screen will look like a busy galaxy of shooting stars, with no larger shapes emerging. But turn up the duration of the trails, and the number of agents, and at a certain clearly defined point, a cluster of cells will suddenly form. The system has entered a phase transition, moving from one discrete state to another, based on the "organized complexity" of the slime mold cells. This is not gradual, but sudden, as though a switch had been flipped. But there are no switch-flippers, no pacemakers—just a swarm of isolated cells colliding with one another, and leaving behind their pheromone footprints.

Histories of intellectual development—the origin and spread of new ideas—usually come in two types of packages: either the "great man" theory, where a single genius has a eureka moment in the lab or the library and the world is immediately transformed; or the "paradigm shift" theory, where the occupants of the halls of science awake to find an entirely new floor has been built on top of them, and within a few years, everyone is working out of the new offices. Both theories are inadequate: the great-man story ignores the distributed, communal effort that goes into any important intellectual advance, and the paradigm-shift model has a hard time explaining how the new floor actually gets built. I suspect Mitch Resnick's slime mold simulation may be a better metaphor for the way idea revolutions come about: think of those slime mold cells as investigators in the field; think of those trails as a kind of institutional memory. With only a few minds exploring a given problem, the cells remain disconnected, meandering across the screen as isolated units, each pursuing its own desultory course. With pheromone trails that evaporate quickly, the cells leave no trace of their progress—like an essay published in a journal that sits unread on a library shelf for years. But plug more minds into the system and give their work a longer, more durable trail—by publishing their ideas in best-selling books, or founding research centers to explore those ideas—and before long the system arrives at a phase transition: isolated hunches and private obsessions coalesce into a new way of looking at the world, shared by thousands of individuals.

This is exactly what happened with the bottom-up mind-set over the past three decades. After years of disconnected investigations, the varied labors of Turing, Shannon, Wiener, Selfridge, Weaver, Jacobs, Holland, and Prigogine had started a revolution in the way we thought about the world and its systems. By the time Jefferson and Taylor started tinkering with their virtual ants in the mideighties, the trails of intellectual inquiry had grown long and interconnected enough to create a higher-level order. (Call it the emergence of emergence.) A field of research that had been characterized by a handful of early-stage investigations blossomed overnight into a densely populated and diverse landscape, transforming dozens of existing disciplines and inventing a handful of new ones. In 1969, Marvin Minsky and Seymour Papert published "Perceptrons," which built on Selfridge's Pandemonium device for distributed pattern recognition, leading the way for Minsky's bottom-up Society of Mind theory developed over the following decade. In 1972, a Rockefeller University professor named Gerald Edelman won the Nobel prize for his work decoding the language of antibody molecules, leading the way for an understanding of the immune system as a self-learning pattern-recognition device. Prigogine's Nobel followed five years later. At the end of the decade, Douglas Hofstadter published *Gödel, Escher, Bach,* linking artificial intelligence, pattern recognition, ant colonies, and "The Goldberg Variations." Despite its arcane subject matter and convoluted rhetorical structure, the book became a best-seller and won the Pulitzer prize for nonfiction.

By the mideighties, the revolution was in full swing. The Santa Fe Institute was founded in 1984; James Gleick's book *Chaos* arrived three years later to worldwide adulation, quickly followed by two popular-science books each called *Complexity.* Artificial-life studies flourished, partially thanks to the success of software programs like Tracker. In the humanities, critical theorists such as Manuel De Landa started dabbling with the conceptual tools of self-organization, abandoning the then-trendy paradigm of poststructuralism or cultural studies. The phase transition was complete; Warren Weaver's call for the study of organized complexity had been vigorously answered. Warren Weavers's "middle region" had at last been occupied by the scientific vanguard.

We are now living through the third phase of that revolution. You can date it back to the day in the early nineties when Will Wright released a program called SimCity, which would go on to become one of the best-selling video-game franchises of all time. SimCity would also inaugurate a new phase in the developing story of self-organizing: emergent behavior was no longer purely an object of study, something to interpret and model in the lab. It was also something you could *build,* something you could interact with, and something you could sell. While SimCity came out of the developing web of the bottom-up worldview, it suggested a whole new opening: SimCity was a work of culture, not science. It aimed to entertain, not explain.

Ten years after Wright's release of SimCity, the world now abounds with these man-made systems: online stores use them to recognize our cultural tastes; artists use them to create a new kind of adaptive cultural form; Web sites use them to regulate their online communities; marketers use them to detect demographic patterns in the general public. The video-game industry itself has exploded in size, surpassing Hollywood in terms of raw sales numbers—with many of the best-selling titles relying on the powers of digital self-organization. And with that popular success has come a subtle, but significant, trickle-down effect: we are starting to *think* using the conceptual tools of bottom-up systems. Just like the clock maker metaphors of the Enlightenment, or the dialectical logic of the nineteenth century, the emergent worldview belongs to this moment in time, shaping our thought habits and coloring our perception of the world. As our everyday life becomes increasingly populated by artificial emergence, we will find ourselves relying more and more on the logic of these systems—both in corporate America, where "bottom-up intelligence" has started to replace "quality management" as the mantra of the day, and in the radical, antiglobalization protest movements, who explicitly model their pacemakerless, distributed organizations after ant colonies and slime molds. Former vice president Al Gore is himself a devotee of complexity theory and can talk for hours about what the bottom-up paradigm could mean for reinventing government. Almost two centuries after Engels wrestled with the haunting of Manchester's city streets, and fifty years after Turing puzzled over the mysteries of a flower's bloom, the circle is finally complete. Our minds may be wired to look for pacemakers, but we are steadily learning how to think from the bottom up.

PART TWO

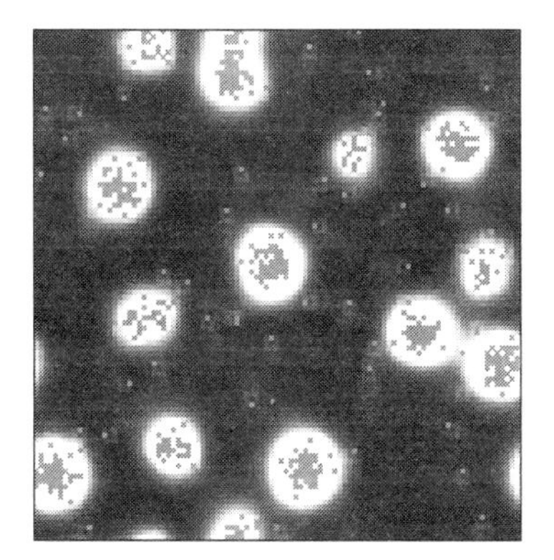

StarLogo slime mold simulation
(Courtesy of Mitch Resnick)

Look to the ant, thou sluggard;
Consider her ways and be wise:
Which having no chief, overseer, or ruler,
Provides her meat in the summer,
And gathers her food in the harvest.

—Proverbs 6:6–8

2

Street Level

Say what you will about global warming or the Mona Lisa, *Apollo 9* or the canals of Venice—human beings may seem at first glance to be the planet's most successful species, but there's a strong case to be made for the ants. Measured by sheer numbers, ants—and other social insects such as termites—dominate the planet in a way that makes human populations look like an evolutionary afterthought. Ants and termites make up 30 percent of the Amazonian rain forest biomass. With nearly ten thousand known species, ants rival modern humans in their global reach: the only large landmasses free of ant natives are Antarctica, Iceland, Greenland, and Polynesia. And while they have yet to invent aerosol spray, ant species have a massive environmental impact, moving immense amounts of soil and distributing nutrients even in the most hostile environments. They lack our advanced forebrains, of course, but human intelligence is only one measure of evolutionary success.

All of which raises the question, if evolution didn't see fit to endow ants with the computational powers of the human brain, how did they become such a dominant presence on the planet? While there's no single key to the success of the social insects, the collective intelligence of the colony system certainly played an essential role. Call it swarm logic: ten thousand ants—each limited to a meager vocabulary of pheromones and minimal cognitive skills—collectively engage in nuanced and improvisational problem-solving. A harvester ant colony in the field will not only ascertain the shortest route to a food source, it will also prioritize food sources, based on their distance and ease of access. In response to changing external conditions, worker ants switch from nest-building to foraging to raising ant pupae. Their knack for engineering and social coordination can be downright spooky—particularly because none of the individual ants is actually "in charge" of the overall operation. It's this connection between micro and macro organization that got Deborah Gordon into ants in the first place. "I was interested in systems where individuals who are unable to assess the global situation still work together in a coordinated way," she says now. "And they manage to do it using only local information."

Local turns out to be the key term in understanding the power of swarm logic. We see emergent behavior in systems like ant colonies when the individual agents in the system pay attention to their immediate neighbors rather than wait for orders from above. They think locally *and* act locally, but their collective action produces global behavior. Take the relationship between foraging and colony size. Harvester ant colonies constantly adjust the number of ants actively foraging for food, based on a number of variables: overall colony size (and thus mouths needed to be fed); amount of food stored in the nest; amount of food available in the surrounding area; even the presence of other colonies in the near vicinity. No individual ant can assess any of these variables on her own. (I use *her* deliberately—all worker ants are females.) The perceptual world of an ant, in other words, is limited to the street level. There are no bird's-eye views of the colony, no ways to perceive the overall system—and indeed, no cognitive apparatus that could make sense of such a view. "Seeing the whole" is both a perceptual and conceptual impossibility for any member of the ant species.

Indeed, in the ant world, it's probably misguided to talk about "views" at all. While some kinds of ants have surprisingly well-developed optical equipment (the South American formicine ant *Gigantiops destructor* has massive eyes), the great bulk of ant information-processing relies on the chemical compounds of pheromones, also known as semiochemicals for the way they create a functional sign system among the ants. Ants secrete a finite number of chemicals from their rectal and sternal glands—and occasionally regurgitate recently digested food—as a means of communicating with other ants. Those chemical signals turn out to be the key to understanding swarm logic. "The sum of the current evidence," E. O. Wilson and Bert Holldobler write in their epic work, *The Ants,* "indicates that pheromones play the central role in the organization of colonies."

Compared to human languages, ant communication can seem crude, typically possessing only ten or twenty signs. Communication between workers in colonies of the fire ant *Solenopsis invicta*—studied intensely by Wilson in the early sixties—relies on a vocabulary of ten signals, nine of which are based on pheromones. (The one exception is tactile communication directly between ants.) Among other things, these semiochemicals code for task-recognition ("I'm on foraging duty"); trail attraction ("There's food over here"); alarm behavior ("Run away!"); and necrophoric behavior ("Let's get rid of these dead comrades").

While the vocabulary is simple, and complex syntactical structures impossible, the language of the ants is nevertheless characterized by some intriguing twists that add to its expressive capability. Many semiochemicals operate in a relatively simple binary fashion—signaling, for instance, whether another ant is a friend or a foe. But ants can also detect *gradients* in pheromones, revealing which way the scent is growing stronger, not unlike the olfactory skills of bloodhounds. Gradient detection is essential for forming those food delivery lines that play such a prominent role in the popular imagination of ant life: the seemingly endless stream of ants, each comically overburdened with seeds, marching steadily across sidewalk or soil. (As we will see in Chapter 5, Mitch Resnick's program StarLogo can also model the way colonies both discover food sources and transport the goods back to the home base.) Gradients in the pheromone trail are the difference between saying "There's food around here somewhere" and "There's food due north of here."

Like most of their relatives, the harvester ants that Deborah Gordon studies are also particularly adept at measuring the *frequency* of certain semiochemicals, a talent that also broadens the semantic range of the ant language. Ants can sense the difference between encountering ten foraging ants in an hour and encountering a hundred. Gordon believes this particular skill is critical to the colony's formidable ability to adjust task allocation according to colony size or food supply—a local talent, in other words, that engenders global behavior.

"I don't think that the ants are assessing the size of the colony," she tells me, "but I think that the colony size affects what an ant experiences, which is different. I don't think that an ant is keeping track of how big the whole colony is, but I think that an ant in a big colony has a different experience from an ant in a small colony. And that may account for why large old colonies act different than their small ones." Ants, in Gordon's view, conduct a kind of statistical sample of the overall population size, based on their random encounters with other ants. A foraging ant might expect to meet three other foragers per minute—if she encounters more than three, she might follow a rule that has her return to the nest. Because larger, older colonies produce more foragers, ants may behave differently in larger colonies because they are more likely to encounter other ants.

This local feedback may well prove to be the secret to the ant world's decentralized planning. Individual ants have no way of knowing how many foragers or nest-builders or trash collectors are on duty at any given time, but they can keep track of how many members of each group they've stumbled across in their daily travels. Based on that information—both the pheromone signal itself, and its frequency over time—they can adjust their own behavior accordingly. The colonies take a problem that human societies might solve with a command system (some kind of broadcast from mission control announcing that there are too many foragers) and instead solve it using statistical probabilities. Given enough ants moving randomly through a finite space, the colony will be able to make an accurate estimate of the overall need for foragers or nest-builders. Of course, it's always possible that an individual ant might randomly stumble across a disproportionate number of foragers and thus overestimate the global foraging state and change her behavior accordingly. But because the decision-making process is spread out over thousands of individuals, the margin of error is vanishingly small. For every ant that happens to overestimate the number of foragers on duty, there's one that underestimates. With a large enough colony, the two will eventually cancel each other out, and an accurate reading will emerge.

If you're building a system designed to learn from the ground level, a system where macrointelligence and adaptability derive from local knowledge, there are five fundamental principles you need to follow. Gordon's harvester ants showcase all of them at work:

More is different. This old slogan of complexity theory actually has two meanings that are relevant to our ant colonies. First, the statistical nature of ant interaction demands that there be a critical mass of ants for the colony to make intelligent assessments of its global state. Ten ants roaming across the desert floor will not be able to accurately judge the overall need for foragers or nest-builders, but two thousand will do the job admirably. "More is different" also applies to the distinction between micromotives and macrobehavior: individual ants don't "know" that they're prioritizing pathways between different food sources when they lay down a pheromone gradient near a pile of nutritious seeds. In fact, if we only studied individual ants in isolation, we'd have no way of knowing that those chemical secretions were part of an overall effort to create a mass distribution line, carrying comparatively huge quantities of food back to the nest. It's only by observing the entire system at work that the global behavior becomes apparent.

Ignorance is useful. The simplicity of the ant language—and the relative stupidity of the individual ants—is, as the computer programmers say, a feature not a bug. Emergent systems can grow unwieldy when their component parts become excessively complicated. Better to build a densely interconnected system with simple elements, and let the more sophisticated behavior trickle up. (That's one reason why computer chips traffic in the streamlined language of zeros and ones.) Having individual agents capable of directly assessing the overall state of the system can be a real liability in swarm logic, for the same reason that you don't want one of the neurons in your brain to suddenly become sentient.

Encourage random encounters. Decentralized systems such as ant colonies rely heavily on the random interactions of ants exploring a given space without any predefined orders. Their encounters with other ants are individually arbitrary, but because there are so many individuals in the system, those encounters eventually allow the individuals to gauge and alter the macrostate of the system itself. Without those haphazard encounters, the colony wouldn't be capable of stumbling across new food sources or of adapting to new environmental conditions.

Look for patterns in the signs. While the ants don't need an extensive vocabulary and are incapable of syntactical formulations, they do rely heavily on patterns in the semiochemicals they detect. A gradient in a pheromone trail leads them toward a food source, while encountering a high ratio of nest-builders to foragers encourages them to switch tasks. This knack for pattern detection allows meta-information to circulate through the colony mind: signs about signs. Smelling the pheromones of a single forager ant means little, but smelling the pheromones of fifty foragers in the space of an hour imparts information about the global state of the colony.

Pay attention to your neighbors. This may well be the most important lesson that the ants have to give us, and the one with the most far-reaching consequences. You can restate it as "Local information can lead to global wisdom." The primary mechanism of swarm logic is the interaction between neighboring ants in the field: ants stumbling across each other, or each other's pheromone trails, while patrolling the area around the nest. Adding ants to the overall system will generate more interactions between neighbors and will consequently enable the colony itself to solve problems and regulate itself more effectively. Without neighboring ants stumbling across one another, colonies would be just a senseless assemblage of individual organisms—a swarm without logic.

Gordon's harvester ant colonies contain another mystery. If we understand how local interactions can lead to global problem-solving, we still don't have an answer to the question of how colonies develop over time. This is one of those scientific questions that nobody thought to ask, because the phenomenon had gone unobserved. And that phenomenon had gone unobserved because people had been thinking about ants—and watching ants—using the wrong *scale*. Until recently, entomologists studied colony behavior in snapshots, surveying a given nest for days or months at a time, then moving on to other nests or back to the lab. But successful colonies can live as long as fifteen years—the life span of the egg-laying queen ant, whose demise signals the final death of the colony itself. Entomologists had been looking at individual colonies in the scale of weeks or months. But to understand how colonies develop, you needed to work on the scale of decades.

In the mideighties, when she first began doing fieldwork in Arizona, Gordon made a bold research gamble that turned out, in hindsight, to be brilliant: she decided to track individual colonies year to year, following them through their birth at the end of a successful mating flight all the way to their fifteen-year-old senescence. After a half decade or so in this time-consuming project, the results began to come in, and they were fascinating. Like a stop-motion film of a vine winding its way around a branch, Gordon's research transformed the way that we think about ants by transforming the temporal scale with which we perceived them. The colonies cycled through a clearly defined infancy, adolescence, and mature phase over their fifteen-year existence. "I had never thought about it, or read anything about it, because without long-term data, nobody really knows the ages of their colonies," she says now. "So it wasn't until I had been watching the same colonies year after year, and began to be able to count how old the colonies were, that I could start to see that young colonies were more active." As she continued her observations, a number of differences emerged between colonies of varying ages, differences that were eerily reminiscent of other developmental cycles in the animal kingdom.

For one, younger colonies are more fickle. "I've done experiments that mimic the kinds of changes in environment that a colony usually experiences—say, a change in the availability of food," Gordon tells me. "If I do the same experiment week after week with older colonies, I get the same results: they respond the same way over and over. If we do the same experiment week after week with a younger colony, they'll respond one way this week, and another way next week, so the younger colonies are more sensitive to whatever's different about this week than last week."

"Typical teenagers," I say, laughing.

"Maybe." She smiles. "And the other thing that might be more typical of teenagers would be the difference between older and younger colonies in the ways that they respond to their neighbors. Neighboring harvester ant colonies meet when foragers from the two colonies overlap and search the same places for food. If older colonies meet a neighbor one day, the next day they're more likely to turn and go in the other direction to avoid each other. The younger colonies are much more persistent and aggressive, even though they're smaller. So they meet one day and they'll go right back the next day—even if they have to fight."

The developmental cycles of colonies may be intriguing enough at face value, but consider this additional fact: while the overall colony evolves and adapts over fifteen years, the ants that make up the colony live no longer than twelve months. Indeed, the hapless male ants—who only show up once a year for the mating flight—only live for a single day. (Their life span is so abbreviated that natural selection didn't bother to endow them with jaws to eat, since they don't live long enough to get hungry.) Only the queen ant lasts for more than a year, and yet she does nothing but lay eggs and is entirely uninvolved with the behavior of worker ants out in the field. The colony grows more stable and less impetuous as it develops, and yet the population of the colony starts over from scratch each year. How does the whole develop a life cycle when the parts are so short-lived?

It would not be wrong to say that understanding emergence begins with unraveling this puzzle. The persistence of the whole over time—the global behavior that outlasts any of its component parts—is one of the defining characteristics of complex systems. Generations of ants come and go, and yet the colony itself matures, grows more stable, more organized. The mind naturally boggles at this mix of permanence and instability. We can understand it when we stumble across, say, a Tudor house in the Cotswolds whose every plank and beam and brick has been replaced at least once in its lifetime, because those bricks are being replaced by "master planners": craftsmen or residents who know what the house itself is supposed to look like, and who deliberately follow the original blueprints. Gordon's ant colonies are more like a house that automatically replaces its skin once a year, without anybody helping out. Or better yet, given that ant colonies grow more durable over time, it's like a house that spontaneously develops a sturdier insulation system after five years and sprouts a new garage after ten.

The ant colony may amaze us with its capacity to grow and evolve while discarding entire generations of worker ants, but as it turns out, we're not all that different from social insects like ants, termites, or bees. As the science writer Matt Ridley observes, "The relationship between body cells is indeed very much like that between bees in a hive. The ancestors of your cells were once individual entities, and their evolutionary 'decision' to cooperate, some six hundred million years ago, is almost exactly equivalent to the same decision, taken perhaps fifty million years ago by the social insects, to cooperate on the level of the body; close genetic relatives discovered they could reproduce more effectively if they did so vicariously, delegating the task to germ cells in the cells' case, or to a queen, in the case of bees."

The human body is made up of several hundred different types of cells—muscle, blood, nervous, and so on. At any given time, approximately 75 trillion of these cells are working away in your body. In a very real sense, you are the sum of their actions; there is no *you* without them. And yet those cells are dying all the time! Thousands probably died in the time it took you to read the last sentence, and by next week, you will be composed of billions of new cells that weren't there to enjoy the reading of that sentence, much less enjoy your first step or your high school prom. Cells are dying all the time in your body—and most of them are being replaced at a tremendous clip. (Even brain cells turn out to regenerate themselves far into adulthood.) And yet somehow, despite that enormous cellular turnover, you still feel like yourself week to week and year to year. How is this possible?

Some readers might be inclined to object at this point that humans are in fact closer to that endlessly rebuilt Tudor house than an ant colony, because in the case of human development we do have a master planner and a blueprint that we can follow: those coils of DNA wrapped neatly in every cell in our body. Our cells know how to build our bodies because natural selection has endowed them with a meticulously detailed plan, and has seen to it that 75 trillion copies are distributed throughout our bodies at any given time. The tyranny of DNA would seem to run counter to the principles of emergence: if all the cells are reading from the same playbook, it's not a bottom-up system at all; it's the ultimate in centralization. It would be like an ant colony where each ant started the day with a carefully planned agenda: forage from six to ten; midden duty until noon; lunch; and then cleanup in the afternoon. That's a command economy, not a bottom-up system.

So does this mean our genes are secret Stalins, doling out the fixed plan for growth to the Stakhanovites of our cells? Are we more like a socialist housing complex than an ant colony? No one questions that DNA exerts an extraordinary influence over the development of our cells, and that each cell in our body contains the same genetic blueprint. If each cell were simply reading from the chromosomal playbook and behaving accordingly, you could indeed make the argument that our bodies don't function like ant colonies. But cells do more than just follow the dictates of DNA. They also learn from their neighbors. And without that local interaction, the master plan of our genetic code would be utterly useless.

Cells draw selectively upon the blueprint of DNA: each cell nucleus contains the entire genome for the organism, but only a tiny segment of that data is read by each individual cell: muscle cells read from the lines of code that concern muscle cells, while blood cells consult the passages that relate to blood cells. This seems simple enough, until you ask the question, how did a muscle cell get to be a muscle cell in the first place? And that question underlies one of the most fundamental mysteries of emergence, which is how complicated organisms, with a wide variety of building blocks, can develop out of such simple beginnings. We all start life as a single-celled organism, and yet by the end of our development cycle, we're somehow composed of two hundred variations, all intricately connected to one another, and all performing stunningly complex tasks. How does an egg somehow know how to build a chicken?

The answer is not all that different from the solution that ant colonies rely on. Cells self-organize into more complicated structures by learning from their neighbors. Each cell in your body contains an intricate set of tools for detecting the state of surrounding cells, and for communicating to those cells using various chemical messengers. Where ants used pheromones to inform each other of their activities, cells communicate via salts, sugars, amino acids—even larger molecules such as proteins and nucleic acids. The messages are partially transmitted through cell "junctions," small passageways that admit molecules from one cell's cytoplasm to another. This communication plays an essential role in all cellular activity, but it is particularly critical for embryonic development during which a single-celled organism self-organizes into a mouse or a roundworm or a human being.

We all begin life as a single-celled embryo, but seconds after conception, the embryo divides itself into two compartments: a "head" and a "tail." At that point, the organism has joined the ranks of multicellular life, being composed now of two distinct cells. And those two cells—the head and the tail—have separate instructions for growth encoded in their DNA: one cell turns to the "head cell" chapter, the other to the "tail cell" chapter. At this early stage of development, the instructions follow a predictable pattern: divide into another "head" and "tail." Thus, in the second round of embryonic development, there are four cells: the head of the head, the tail of the head, the head of the tail, and the tail of the tail. Those four units may not sound like much, but this cycle of cell division continues at a blistering clip. A frog embryo self-divides into nearly ten thousand cells in a matter of hours. The runaway power of geometric progression is not just a mathematical oddity—it is also essential to the very origins of life.

Once the embryo reaches a certain size, cell "collectives" start to form, and here matters get more complicated. One group of cells may be the beginning of an arm, while another group may be the first stirrings of the brain's gray matter. Each cell has somehow to figure out where it is in the larger scheme of things—and yet, like the ants, cells have no way of seeing the whole, and they have no fixed address stamped upon them when they come into the world, no factory serial number. But while cells lack a bird's-eye view of the organism that contains them, they can make street-level assessments via the molecular signals transmitted through the cell junctions. This is the secret of self-assembly: cell collectives emerge because each cell looks to its neighbors for cues about how to behave. Those cues directly control what biologists call "gene expression"; they're the cheat sheet that enables each cell to figure out which segment of DNA to consult for its instructions. It's a kind of microscopic herd mentality: a cell looks around to its neighbors and finds that they're all working away steadily at creating an eardrum or a heart valve, which in turn causes the cell to start laboring away at the same task.

The key here is that life does not simply reduce down to transcribing static passages from our genetic scripture. Cells figure out which passages to pay attention to by observing signals from the cells around them: only with that local interaction can complex "neighborhoods" of cell types come into being. The Nobel laureate Gerald Edelman calls this process topobiology, from the Greek word for "place," *topos.* Cells rely heavily on the code of DNA for development, but they also need a sense of place to do their work. Indeed, the code is utterly worthless without the cell's ability to determine its place in the overall organism, a feat that is accomplished by the elegant strategy of paying attention to one's neighbors. As Ridley writes, "The great beauty of embryo development, the bit that human beings find so hard to grasp, is that it is a totally decentralized process. Since every cell in the body carries a complete copy of the genome, no cell need wait for instructions from authority; every cell can act on its own information and the signals it receives from its neighbors." And so we have come full circle back to Gordon's ants, and their uncanny ability to generate coordinated global behavior out of local interactions.

* * *

Neighbors and *neighborhoods*. The words seem more attached to the communities of human settlements than the microscopic domains of muscle cells or harvester ants. But how do we extend our vision up one more level on the chain of life to the cultural "superorganism" of the city? Certainly it is possible to *model* the behavior of cities by using the tools of swarm logic. Computer-based simulations can teach us a tremendous amount about complex systems: if a picture is worth a thousand words, an interactive model must be valued in the millions. But a quick look at the software best-seller lists will tell you that city simulations are more than just an educational device. Will Wright's SimCity franchise has now sold millions of copies; it's likely that the number of virtual towns created using Wright's tools exceeds the number of real towns formed in modern human history. Some games attract our attention by appealing to our appetite for storytelling, following a linear progression of move and countermove, with clearly defined beginnings and endings; other games catch the eye by blowing things up. SimCity was one of the first games to exploit the uncanny, bottom-up powers of emergence. Wright's genius was not simply in recognizing the fun of simulating an entire metropolis on your screen. He also hit upon a brilliant programming trick that enabled the city to evolve in a more lifelike way—a trick that closely resembles the behavior of ant colonies and embryos.

Much has been made of the fact that you can't ever "win" at SimCity, but it's probably more important to note that you don't really "play" SimCity either, at least the way we talk about playing conventional games. Users *grow* their virtual cities, but the cities evolve in unpredictable ways, and control over the city's eventual shape is always indirect. You can create commercial zones or build a highway, but there's never a guarantee that the neighborhood will take off or the crime rate go down. (It's far from random, of course—longtime players learn how to push their virtual citizens in certain directions.) For most people, the sight of their first digital town sprouting upscale neighborhoods and chronically depressed slums is downright eerie, as though the hard math of the digital computer had somehow generated a life-form, something more organic and fluid, somewhere between the rigid dictates of programming and pure randomness.

How did Wright create this extraordinary illusion? By designing the game as an emergent system, a meshwork of cells that are connected to other cells, and that alter their behavior in response to the behavior of other cells in the network. A given city block in SimCity possesses a number of values—the price of the land, say, or its pollution level. As in a real-world city, these values change in response to the values of neighboring blocks; if the block to the west drops in value, and the eastern neighbor develops a higher crime rate, then the current block may well grow a little less valuable. (A sophisticated SimCity player might counter the decline by placing a police station within ten blocks of the depressed area.) The algorithms themselves are relatively simple—look at your neighbors' state, and change your state accordingly—but the magic of the simulation occurs because the computer makes thousands of these calculations per second. Because each cell is influencing the behavior of other cells, changes appear to ripple through the entire system with a fluidity and definition that can only be described as lifelike.

The resemblance to our ants and embryos is striking. Each block in SimCity obeys a set of rigid instructions governing its behavior, just as our cells consult the cheat sheet of our genes. But those instructions are dependent on the signals received from other blocks in the neighborhood, just as cells peer out through gap junctions to gauge the state of their neighbors. With only a handful of city blocks, the game is deathly boring and unconvincingly robotic. But with thousands of blocks, each responding to dozens of variables, the simulated cityscape comes to life, sprouting upscale boroughs and slums, besieged by virtual recessions and lifted by sudden booms. As with ant colonies, more is different. "Great cities are not like towns only larger," Jane Jabobs writes. "They are not like suburbs only denser. They differ from towns and suburbs in basic ways." She was writing, of course, about real-world cities, but she could just as easily have been talking about SimCity's networked algorithms, or the teeming colonies of Arizona harvester ants.

Economists and urban sociologists have also been experimenting with models that can simulate the ways that cities self-organize themselves over time. While actual cities are heavily shaped by top-down forces, such as zoning laws and planning commissions, scholars have long recognized that bottom-up forces play a critical role in city formation, creating distinct neighborhoods and other unplanned demographic clusters. In recent years, some of those theorists—not to mention a handful of mainstream economists—have developed more precise models that re-create the neighborhood-formation process with startling precision.

The economist (and now *New York Times* editorialist) Paul Krugman's 1995 lectures, "The Self-Organizing Economy"—published as a book the following year—include a remarkably simple mathematical model that can account for the "polycentric, plum-pudding pattern of the modern metropolis." Building on the game-theory models that Thomas Schelling developed to explain how segregated cities can form, Krugman's system assumes a simplified city made up only of businesses, each of which makes a decision about where to locate itself based on the location of other businesses. Some centripetal forces draw businesses closer to one another (because firms may want to share a customer base or other local services), and some centrifugal forces drive businesses farther apart (because firms compete for labor, land, and in some cases customers). Within that environment, Krugman's model relies on two primary axioms:

1. There must be a tension between centripetal and centrifugal forces, with neither too strong.
2. The range of the centripetal forces must be shorter than that of the centrifugal forces: business must like to have other businesses nearby, but dislike having them a little way away. (A specialty store likes it when other stores move into its shopping mall, because they pull in more potential customers; it does not like it when stores move into a rival mall ten miles away.)

"And that's all that we need," Krugman continues. "In any model meeting these criteria, any initial distribution of businesses across the landscape, no matter how even (or random), will spontaneously organize itself into a pattern with multiple, clearly separated business centers."

Krugman even provides a chart demonstrating the city's self-organization in time—an image that captures the elegance of the model. Scatter a thousand businesses across this landscape at random, then turn on the clock and watch them shuffle around the space. Eventually, no matter what the initial configuration, the firms will gather into a series of distinct clusters evenly spaced from each other. There's no rule for clustering that the businesses are directly obeying: their motives are strictly local. But those micromotives nevertheless combine to form macrobehavior, a higher order that exists on the level of the city itself. Local rules lead to global structure—but a structure that you wouldn't necessarily predict from the rules.

Krugman talks about his "plum pudding" polycentrism as a feature of the modern "edge city," but his model might also explain an older convention: the formation of neighborhoods within a larger metropolitan unit. Neighborhoods are themselves polycentric structures, born of thousands of local interactions, shapes forming within the city's larger shape. Like Gordon's ant colonies, or the cells of a developing embryo, neighborhoods are patterns in time. No one wills them into existence single-handedly; they emerge by a kind of tacit consensus: the artists go here, the investment bankers here, Mexican-Americans here, gays and lesbians here. The great preponderance of city dwellers live by those laws, without any legal authority mandating that compliance. It is the sidewalk—the public space where interactions between neighbors are the most expressive and the most frequent—that helps us create those laws. In the popular democracy of neighborhood formation, we vote with our feet.

A friend of mine who moved to California a few years ago once remarked to me, with a straight face, "The class segregation in Los Angeles is not nearly as bad as you might think. You'd be surprised how many low-income areas I pass on the freeway when I'm driving into work."

It was one of those comments that reveals an entire weltanschauung. "It's not 'an encounter with the working class,'" I thundered back, "if you're gazing down at them from the overpass." But he had a point. In a dispersed, car-centric city like Los Angeles, highways are the connecting nodes, one of the few zones where the city's different groups encounter each other—albeit at sixty-five miles an hour.

Ever since *Death and Life* was first published in the early sixties, Jacobs-inspired critics have lambasted the dispersed communities of L.A. and Phoenix, and their even more anonymous descendants—the "edge cities" that have sprouted up around convenient freeway intersections or high-volume parking lots, the way towns once nestled up to harbors or major rivers. Progressive urbanists bemoaned the mallification of the American city, with vibrant public streets giving way to generic, private shopping complexes. The sidewalk carnivalesque that had so vividly been captured by Wordsworth and Baudelaire in the previous century seemed headed the way of the horse and buggy, and in each case, the culprit turned out to be the same: the automobile, which necessitated all the injuries of sprawl—mixed-use zoning, gated communities, deserted or nonexistent sidewalks.

At the core of this lamentable transformation was the street itself, and the interactions between strangers that once took place on it. The brilliance of *Death and Life* was that Jacobs understood—before the sciences had even developed a vocabulary to describe it—that those interactions enabled cities to create emergent systems. She fought so passionately against urban planning that got people "off the streets" because she recognized that both the order and the vitality of working cities came from the loose, improvised assemblages of individuals who inhabited those streets. Cities, Jacobs understood, were created not by central planning commissions, but by the low-level actions of borderline strangers going about their business in public life. Metropolitan space may habitually be pictured in the form of skylines, but the real magic of city living comes from below.

Part of that magic is the elemental human need of safety. Chapter 2 of *Death and Life* investigates the way dense urban settlements collectively "solve" the problem of making themselves safe, a solution that has everything to do with the local interactions of strangers sharing the public space of the sidewalks:

> Under the seeming disorder of the old city, wherever the old city is working successfully, is a marvelous order for maintaining the safety of the streets and the freedom of the city. It is a complex order. Its essence is intimacy of sidewalk use, bringing with it a

> constant succession of eyes. This order is all composed of movement and change. . . . The ballet of the good city sidewalk never repeats itself from place to place to place, and in any one place is always replete with new improvisations.

After a long and wonderfully detailed portrait of one day's choreography, Jacobs ends with one of the great passages in the history of cultural criticism:

> I have made the daily ballet of Hudson Street sound more frenetic than it is, because writing it telescopes it. In real life, it is not that way. In real life, to be sure, something is always going on, the ballet is never at a halt, but the general effect is peaceful and the general tenor even leisurely. People who know well such animated city streets will know how it is. I am afraid people who do not will always have it a little wrong in their heads—like the old prints of rhinoceroses made from travelers' descriptions of the rhinoceroses.
>
> On Hudson Street, the same as in the North End of Boston or in any other animated neighborhoods of great cities, we are not innately more competent at keeping the sidewalks safe than are the people who try to live off the hostile truce of turf in a blind-eyed city. We are the lucky possessors of a city order that makes it relatively simple to keep the peace because there are plenty of eyes on the street. But there is nothing simple about that order itself, or the bewildering number of components that go into it. Most of those components are specialized in one way or another. They unite in their joint effect upon the sidewalk which is not specialized in the least. That is its strength.

Again, we are back to the world of the ants: random local interactions leading to global order; specialized components creating an unspecialized intelligence; neighborhoods of individuals solving problems without any of those individuals realizing it. And safety is only part of the story: there are many "uses of sidewalks" in *Death and Life*, some of which we will encounter in later chapters.

The key here is that sidewalks are important *not* because they provide an environmentally sound alternative to freeways (though that is also the case) nor because walking is better exercise than driving (though that too is the case) nor because there's something quaintly old-fashioned about pedestrian-centered towns (that is more a matter of fashion than empirical evidence). In fact, there's nothing about the physical existence of sidewalks that matters to Jacobs. What matters is that they are the primary conduit for the flow of information between city residents. Neighbors learn from each other because they pass each other—and each other's stores and dwellings—on the sidewalk. Sidewalks allow relatively high-bandwidth communication between total strangers, and they mix large numbers of individuals in random configurations. Without the sidewalks, cities would be like ants without a sense of smell, or a colony with too few worker ants. Sidewalks provide both the right *kind* and the right *number* of local interactions. They are the gap junctions of city life.

This is one of those instances where thinking about a social problem using the conceptual tools of emergence sheds genuinely new light on the problem, and on the ways it has been approached in the past. Since *Death and Life*, the celebration of sidewalk culture has become an idée fixe of all left-leaning urbanists, an axiom as widely agreed upon as any in the liberal canon. But the irony is that many of the same critics who cited Jacobs as the initial warrior in the sidewalk crusade misunderstood the reasons why she had embraced the sidewalk in the first place. And that is because they saw the city as a kind of political theater, and not as an emergent system. The clash and contradiction of city streets—versus the antiseptic segregations of suburbia—became a virtue in and of itself, something that people should be "exposed to" for their own good. The logic was a kind of inverted rendition of the old bromides about kids watching too much television: if people were somehow deprived of the theatrical conflicts of city sidewalks, they'd all end up hollow men—or worse, Republicans.

This turns out to be an aesthetic agenda wrapped up in a thin veil of politics. Some critics carried their paeans to sidewalk diversity to laughably condescending extremes. "Poor people have taught us so much about what we know about being fully alive in public," Marshall Berman wrote in an early-eighties essay called "Take It to the Streets." "[They've taught us] about how to move rhythmically and melodically down a street; about how to use color and ornamentation to say new things about ourselves, and to make new connections with the world; about how to bring out the rhetorical and theatrical powers of the English language in our everyday talk." Paraphrase: Those poor people have so much rhythm!

However much Berman might resist the idea, the very same morality play underlies my friend's ode to L.A. freeway culture: both perspectives assume that *seeing* racial and economic diversity is intrinsically good for you, like some kind of political cardiovascular workout. From this perspective, what was laughable about my friend's observation was the idea that he could truly take in the "melodic movements" or hear the "rhetorical" flourishes of South Central while driving on the highway. The exposure itself is assumed prima facie to be good for the soul. The only question is whether my friend was getting a big enough dosage from his car.

This is all perfectly commendable, if a little patronizing, and for all I know we might indeed turn out to be more charitable and expansive people if we encountered more diversity on our streets. But that diet has nothing to do with the Jacobs understanding of sidewalks and their uses. According to the gospel of *Death and Life*, individuals only benefit *indirectly* from their sidewalk rituals: better sidewalks make better cities, which in turn improve the lives of the city dwellers. The value of the exchange between strangers lies in what it does for the superorganism of the city, not in what it does for the strangers themselves. The sidewalks exist to create the "complex order" of the city, not to make the citizens more well-rounded. Sidewalks work because they permit local interactions to create global order.

From this angle, then, the problem with my friend's sojourns on the Santa Monica Freeway—and indeed the problem with all car-centric cities—is that the potential for local interaction is so limited by the speed and the distance of the automobile that no higher-level order can emerge. For all we know, there may well be something psychologically broadening in gazing out over the slums from your Ford Explorer, but that experience will do nothing for the larger health of the city itself, because the information transmitted between agents is so famished and so fleeting. City life depends on the odd interaction between strangers that changes one individual's behavior: the sudden swerve into the boutique you've never noticed before, or the decision to move out of the neighborhood after you pass the hundredth dot-com kid on a cell phone. Encountering diversity does nothing for the global system of the city unless that encounter has a chance of altering your behavior. There has to be feedback between agents, cells that change in response to the changes in other cells. At sixty-five miles an hour, the information transmitted between agents is too limited for such subtle interactions, just as it would be in the ant world if a worker ant suddenly began to hurtle across the desert floor at ten times the speed of her neighbors.

And so this is the ultimate lesson of Jacobs's sidewalks, and of her way of thinking about cities as self-organizing systems. The information networks of sidewalk life are fine-grained enough to permit higher-level learning to emerge. The cars occupy a different scale from the sidewalks, and so the lines of communication between the two orders are necessarily finite. At highway speed, the only complex systems that form are between the cars themselves—in other words, between agents that operate on the same scale. Unlike the ballet of the pedestrian city, these are global patterns that would be familiar to any resident of Los Angeles. We call them traffic jams.

An important distinction must be drawn between ant colonies and cities, though, and it revolves around the question of volition. In a harvester ant colony, the individual ants are relatively stupid, following elemental laws without anything resembling free will. As we have seen, the intelligence of the colony actually relies on the stupidity of its component parts: an ant that suddenly started to make conscious decisions about, say, the number of ants on midden duty would be disastrous for the overall group. You can make the case that this scenario doesn't apply at all to human settlements: cities are higher-level organisms, but their component parts—humans—are far more intelligent, and more self-reflective, than ants are. We consciously make decisions about where to live or shop or stroll; we're not simply driven by genes and pheromones. And so the social patterns we form tend to be substantially more complex than those of the ant world.

Even Gordon herself is sympathetic to the objection. "In a human society, every person always thinks they know what they're doing, even if they're wrong," she says to me near the end of my visit. "It's very hard to imagine any human society in which people would go around responding to what happened at the moment without any conception of why they're doing what they're doing.

That's why I'm always hesitant to make analogies from ants to people, because ants are so unlike people. In fact I think it's the alienness of ants that makes them so intriguing."

Gordon's caveats are important, and as we have already seen, cities involve countless elements that are the exact opposites of those bottom-up systems. (Even SimCity has a mayor!) But the fact that humans think for themselves, and the fact that city organization relies on both hierarchies and heterarchies, does not mean that Wordsworth's "ant-hill on the plain" belongs purely to the world of metaphor. Certain key elements of traditional urban life—indeed, some of the elements that we most cherish about our cities—belong squarely to the world of emergence. What ants do and what cells do and what sidewalks do should be seen as instances of the same idea, the same activity built out of varied material, like a musical score played by different instruments. But to see beyond the objections of individual human volition, we need to think about cities on the right scale. The emphasis on free will only matters on the scale of the individual human life. We need to think about cities the way Gordon thought about ant colonies—on the scale of the superorganism itself.

The decision-making of an ant exists on a minute-by-minute scale: counting foragers, following pheromone gradients. The sum of all those isolated decisions creates the far longer lifetime of the colony, but the ants themselves are utterly ignorant of that macrolevel. Human behavior works at two comparable scales: our day-to-day survival, which involves assessments of the next thirty or forty years at best; and the millennial scale of cities and other economic ecosystems. Driving a car has short-term and long-term consequences. The short term influences whether we make it to soccer practice on time; the long term alters the shape of the city itself. We interact directly with, take account of—and would seem to *control*—the former. We are woefully unaware of the latter. Our decisions to shop at a local boutique or move from one neighborhood to another or even leave the city altogether are all made on the scale of the human lifetime—and usually a much shorter time frame than that. Those decisions we make consciously, but they also contribute to a macrodevelopment that we have almost no way of comprehending, despite our advanced forebrains. And that macrodevelopment belongs to the organism of the city itself, which grows and evolves and learns over a thousand-year cycle, as dozens of human generations come and go.

Viewed at that speed—the millennium's time-lapse footage—our individual volition doesn't seem all that different from that of Gordon's harvester ants, each of whom only lives to see a small fraction of the colony's fifteen-year existence. Those of us who walk the sidewalks of today's cities remain as ignorant of the long-term view, the thousand-year scale of the metropolis, as the ants are of the colony's life. Perceived at that scale, the success of the urban superorganism might well be the single most momentous global event of the past few centuries: until the modern era less than 3 percent of the world's population lived in communities of more than five thousand people; today, half the planet lives in urban environments. Just as the social insects deserve to be seen as some of the planet's most successful organisms, so too should the superorganism of the city; not necessarily because cities are more humane or civilized places, but because they have done such a good job of replicating themselves, drawing in migrant populations from around the world, and encouraging—for the most part—higher birth rates and longer life spans within their confines. You can debate the merits of the transformation, but the fact is that human life on earth now unfolds in cities more often than not. Quantitatively, we are a species of city dwellers now.

Why has the city superorganism triumphed over other social forms? As in the case of the social insects, there are a number of factors, but a crucial one is that cities, like ant colonies, possess a kind of emergent intelligence: an ability to store and retrieve information, to recognize and respond to patterns in human behavior. We contribute to that emergent intelligence, but it is almost impossible for us to perceive that contribution, because our lives unfold on the wrong scale. The next chapter is an attempt to see our way around that blind spot.

3

The Pattern Match

In the final decades of the twelfth century, the Societas Mercatorum, the organization of merchants that had presided over the commercial culture of Florence for nearly a hundred years, began to break apart into splinter groups: guilds with names like the Arte di Por Santa Maria and the Arte di Calimala, structured around specific trades—blacksmiths, moneylenders, wine merchants. A few guilds incorporated diverse groups under one umbrella. One such guild, the Arte di Por Santa Maria, included both silk weavers and goldsmiths.

The creation of the guild system, by all accounts, proved to be a reorganization that literally changed the world. Historians like to talk up the aesthetic accomplishments of the Renaissance, but the guild system pioneered in Florence had as much of an impact on Western civilization as anything dreamed up by da Vinci or Brunelleschi. The gold florin, the local coin minted by the Florentine guilds, was for a long stretch the standard currency of Europe, and one of the first since Roman days to be honored so widely. A number of inventions that turned out to be essential to modern commercial life—double-entry accounting, to name one—date back to the golden age of the guilds. If the engine of history restarted in Italy during the twelfth and thirteenth centuries, as the canonical story goes, the guilds were its turbines.

The guild of Por Santa Maria took its name from a central street that leads directly to the ancient Ponte Vecchio, the much-photographed bridge spanning the River Arno, overloaded with shops and a secret corridor built for the Florentine duke Cosimo I in 1565. There are records of silk weavers setting up shop along the Por Santa Maria as early as 1100, a century before joining forces with the goldsmiths to form their own guild. Merchants who were in the silk trade and other wealthy Florentines could stroll down to the Por Santa Maria comparison shopping, while their servants combed the Ponte Vecchio for the meat sold by the butchers who populated the bridge for the first centuries of the millennium.

They are still there today. Walk north of the Ponte Vecchio on a weekday morning, and you'll still find stores selling fine silks, some of them hawking processed items such as blouses and scarves, others selling the raw goods directly, as they did nearly a millennium ago.

Do cities learn? Not the individuals who populate cities, not the institutions they foster, but the cities themselves. I think the answer is yes. And the silk weavers of Florence can help explain why.

Learning is one of those activities that we habitually associate with conscious awareness—such as falling in love or mourning the loss of a relative. But learning is a complicated phenomenon that exists on a number of levels simultaneously. When we say we "learn someone's face," there's a strong implication of consciousness in the statement—you *feel* something different when you see someone you know, and that feeling of recognition is part of what it means to learn, so much so that it can sometimes seem interchangeable with the experience itself.

But learning is not always contingent on consciousness. Our immune systems learn throughout our lifetimes, building vocabularies of antibodies that evolve in response to the threat posed by invading microorganisms. Most of us have developed immunity to the varicella-zoster virus—also known as the chicken pox—based on our exposure to it early in childhood. That immunity is a learning process: the antibodies of our immune system learn to neutralize the antigens of the virus, and they remember those neutralization strategies for the rest of our lives. We don't come into the world predisposed to ward off the chicken pox virus—our bodies learn how to do it on the fly, without any specific training. Those antibodies function as a "recognition system," in Gerald Edelman's phrase, successfully attacking the virus and storing the information about it, then recalling that information the next time the virus comes across the radar.

Like a six-month-old infant, the immune system first learns to recognize things that differ from itself, then sets out to control those things. It is only part of the wonder of this process that it works as well as it does. What's equally amazing is the fact that the recognition unfolds purely on a cellular level: we are not *aware* of the varicella-zoster virus in any sense of the word, and while our minds may remember what it was like to have chicken pox as a child, our conscious memory has nothing to do with our resistance to the disease.

The body learns without consciousness, and so do cities, because learning is not just about being *aware* of information; it's also about storing information and knowing where to find it. It's about being able to recognize and respond to changing patterns—the way Oliver Selfridge's Pandemonium software does or Deborah Gordon's harvester ants. It's about altering a system's behavior in response to those patterns in ways that make the system more successful at whatever goal it's pursuing. The system need not be conscious to be capable of that kind of learning, just as your immune system need not be conscious to learn how to protect you from the chicken pox.

Imagine a contemporary citizen of Florence who time-travels back eight hundred years, to the golden age of the guilds. What would that experience—the "shock of the old"—be like? Most of it would be utterly baffling: few of modern Florence's landmarks would exist—the Uffizi, say, or the church of San Lorenzo. Only the baptistery of the Duomo would be recognizable, as would the ancient city hall, the Bargello. The broad outline of most streets would look familiar, but in many cases their names would have changed, and our time traveler would find almost nothing recognizable in the buildings lining those streets. The cultural life of the city would be even more disconcerting: the systems of trade and governance would look nothing like those of present-day Florence. Our time traveler might catch some familiar words in the spoken tongue, since the Italian language is a product of Florentine culture, dating back to the turn of the millennium. But if he traveled anywhere else in Italy, he would face serious linguistic hurdles—until the late thirteenth century, Latin was the only language common to all Italians.

And yet, despite that abject confusion, one extraordinary thing remains constant: our time traveler would still know where to buy a yard of silk. Fast-forward a few hundred years, and he'd know where to pick up a gold bracelet as well. And where to buy leather gloves, or borrow money. He wouldn't be equipped to buy any of these things, or even to communicate intelligibly to the salesmen—but he'd know where to find the goods all the same.

Like any emergent system, a city is a pattern in time. Dozens of generations come and go, conquerors rise and fall, the printing press appears, then the steam engine, then radio, television, the Web—and beneath all that turbulence, a pattern retains its shape: silk weavers clustered along Florence's Por Santa Maria, the Venetian glassblowers on Murano, the Parisian traders gathered in Les Halles. The world convulses, sheds its skin a thousand times, and yet the silk weavers stay in place. We have a tendency to relegate these cross-generational patterns to the ossified nostalgia of "tradition," admiring for purely sentimental reasons the blacksmith who works in the same shop as his late-medieval predecessors. But that continuity has much more than sentimental value, and indeed it is more of an achievement than we might initially think. That pattern in time is one of the small miracles of emergence.

Why do cities keep their shapes? Certain elements of urban life get passed on from generation to generation because they're associated with a physical structure that has its own durability. (Cathedrals and universities are the best examples of this phenomenon—St. Peter's Basilica has fostered a religious-themed neighborhood west of the Tiber for a thousand years, and the Left Bank has been a hotbed of student types since the Sorbonne was founded in 1257.) But because those neighborhoods are anchored by specific structures, their persistence has as much to do with the laws of physics as anything else: as long as the cathedral doesn't burn down or disintegrate, there's likely to be a religious flavor to the streets around it. But the Florentine silk weavers are a different matter. There's nothing in the physical structure of the shops that mandates that they be occupied by silk weavers. (Indeed, many of the buildings along the Por Santa Maria have been rebuilt several times over the past thousand years.) They could just as easily house bankers or wine merchants or countless other craftsmen. And yet the silk weavers remain, held in place by the laws of emergence, by the city's gift for self-organization.

You could argue that the silk weavers stay put not because they are part of an emergent system, but because they are subject to the laws of inertia. They remain clustered along the Por Santa Maria because staying put is easier than moving. (In other words, it's not emergence we're seeing here—it's laziness.) The objection might make some sense if we were talking about a fifty-year span, or even a century. But on a thousand-year scale, the force of cultural drift becomes far more powerful. Technological and geopolitical changes obviously have a tremendous impact—killing off entire industries, triggering mass migrations, launching wars, or precipitating epidemics. Neighborhood clusters are extremely vulnerable to those dramatic forces of change, but they are also vulnerable to the slower, mostly invisible drift that all culture undergoes. Over twenty or thirty generations, even something as fundamental as the name of a common item can be transformed beyond recognition, and the steady but imperceptible shifts in pronunciation can make a spoken language unintelligible to listeners. However difficult it is to read Chaucer's *Canterbury Tales* in the original, it would be even more disorienting to hear it read aloud by an inhabitant of fourteenth-century Britain. And if words can transform themselves over time, the changes in social mores, etiquette, and fashion are so profound as to be almost unimaginable. (Parsing the complex sexual codes of thirteenth-century Florence from a modern perspective would be a daunting task indeed.) Viewed on the scale of the millennium, the values of Florentine society look more like a hurricane than a stable social order: all turbulence and change. And yet against all those disruptive forces, the silk weavers hold their own.

Cities are blessed with an opposing force that keeps the drift and tumult of history at bay: a kind of self-organizing stickiness that allows the silk weavers to stay huddled together along the same road for a thousand years, while the rest of the world reinvents itself again and again. These clusters are like magnets planted in the city's fabric, keeping like minds together, even as the forces of history try to break them apart. They are not limited to Italian cities, though Florence's clusters are some of the most ancient. Think of London's Savile Row or Fleet Street, clusters that date back hundreds of years. In Beijing, street names still echo the pockets of related businesses: Silk-Brocade Hat Alley, Dry-Noodle Street. In Manhattan today you can see the early stirrings of clusters, some of them only a few decades old: the diamond row of West Forty-seventh Street, the button district, even a block downtown devoted solely to restaurant supply stores. The jewelry merchants on West Forty-seventh don't have quite the pedigree of their colleagues on the Ponte Vecchio, but then New York is a young city by Italian standards. Look at those Manhattan streets from the thousand-year view, the scale of the superorganism, and what comes to mind is an embryo self-organizing into recognizable shapes, forming patterns that will last a lifetime.

"From its origins onward," Lewis Mumford writes in his classic work *The City in History*, "the city may be described as a structure specially equipped to store and transmit the goods of civilization." Preeminent among the "goods" stored and transmitted by the city is the invaluable material of information: current prices in the marketplace; laborsaving devices dreamed up by craftsmen; new remedies for disease. This knack for capturing information, and for bringing related pockets of information together, defines how cities learn. Like-minded businesses cluster together because there are financial incentives to do so—what academics call economies of agglomeration—enabling craftsmen to share techniques and services that they wouldn't necessarily be able to enjoy on their own. That clustering becomes a self-perpetuating cycle: potential consumers and employees have an easier time finding the goods and jobs they're looking for; the shared information makes the clustered businesses more competitive than the isolated ones.

There are manifest purposes to a city—reasons for being that its citizens are usually aware of: they come for the protection of the walled city, or the open trade of the marketplace. But cities have a latent purpose as well: to function as information storage and retrieval devices. Cities were creating user-friendly interfaces thousands of years before anyone even dreamed of digital computers. Cities bring minds together and put them into coherent slots. Cobblers gather near other cobblers, and button makers near other button makers. Ideas and goods flow readily within these clusters, leading to productive cross-pollination, ensuring that good ideas don't die out in rural isolation. The power unleashed by this data storage is evident in the earliest large-scale human settlements, located on the Sumerian coast and in the Indus Valley, which date back to 3500 B.C. By some accounts, grain cultivation, the plow, the potter's wheel, the sailboat, the draw loom, copper metallurgy, abstract mathematics, exact astronomical observation, the calendar—all of these inventions appeared within centuries of the original urban populations. It's possible, even likely, that more isolated groups or individuals had stumbled upon some of those technologies at an earlier date, but they didn't become part of the collective intelligence of civilization until there were cities to store and transmit them.

The neighborhood system of the city functions as a kind of user interface for the same reason that traditional computer interfaces do: there are limits to how much information our brains can handle at any given time. We need visual interfaces on our desktop computers because the sheer quantity of information stored on our hard drives—not to mention on the Net itself—greatly exceeds the carrying capacity of the human mind. Cities are a solution to a comparable problem, both on the level of the collective and the individual. Cities store and transmit useful new ideas to the wider population, ensuring that powerful new technologies don't disappear once they've been invented. But the self-organizing clusters of neighborhoods also serve to make cities more intelligible to the individuals who inhabit them—as we saw in the case of our time-traveling Florentine. The specialization of the city makes it smarter, more useful for its inhabitants. And the extraordinary thing again is that this learning emerges without anyone even being aware of it. Information management—subduing the complexity of a large-scale human settlement—is the *latent* purpose of a city, because when cities come into being, their inhabitants are driven by other motives, such as safety or trade. No one founds a city with the explicit intent of storing information more efficiently, or making its social organization more palatable for the limited bandwidth of the human mind. That data management only happens later, as a kind of collective afterthought: yet another macrobehavior that can't be predicted from the micromotives. Cities may function like libraries and interfaces, but they are not built with that explicit aim.

Indeed, traditional cities—like the ones that sprouted across Europe between the twelfth and fourteenth centuries—are rarely built with any aim at all: they just happen. There are exceptions of course: imperial cities, such as St. Petersburg or Washington, D.C., laid out by master planners in the image of the state. But organic cities—Florence or Istanbul or downtown Manhattan—are more an imprint of collective behavior than the work of master planners. They are the sum of thousands of local interactions: clustering, sharing, crowding, trading—all the disparate activities that coalesce into the totality of urban living.

All of which raises the question of why—if they are so useful—cities took so long to emerge, and why history includes such long stretches of urban decline. Consider the state of Europe after the fall of the Roman Empire: for nearly a thousand years, European cities retreated back into castles and fortresses, or scattered their populations across the countryside. Imagine a time-lapse film of western Europe, as seen by a satellite, with each decade compressed down to a single second. Start the film at A.D. 100 and the continent is a hundred points of lights, humming with activity. Rome itself glows far brighter than anything else on the map, but the rest of the continent is dotted with thriving provincial capitals: Córdoba, Marseilles, even Paris is large enough to span the Left Bank. As the tape plays, though, the lights begin to dim: cities sacked by invading nomads from the East, or withered away by the declining trade lines of the Empire itself. The Parisians retreat back to their island fortress and remain there for five hundred years. When the Visigoths finally conquer Rome in 476, the satellite image suggests that the power grid of Europe has lost its primary generator: all the lights fade dramatically, and some go out altogether. The system of Europe shifts from a network of cities and towns to a scattered, unstable mix of hamlets and migrants, with the largest towns holding no more than a thousand inhabitants. It stays that way for five hundred years.

And then, suddenly, just after the turn of the millennium, the picture changes dramatically: the continent sprouts dozens of sizable towns, with populations in the tens of thousands. There are pockets on the map—at Venice or Trieste—that glow almost as brightly as ancient Rome had at the start of the tape, nascent cities supporting more than a hundred thousand citizens. The effect is not unlike watching a time-lapse film of an open field, lying dormant through the winter months, then in one sudden shift bursting with wildflowers. There is nothing gradual or linear about the change; it is as sudden, and as emphatic, as turning on a light switch. As the physicist Arthur Iberall once described the process, Europe underwent a transition not unlike that between H_2O molecules changing from the fluid state of water to the crystallized state of ice: for centuries the population is liquid and unsettled—and then, suddenly, a network of towns comes into existence, possessing a stable structure that would persist more or less intact until the next great transformation in the nineteenth century, during the rise of the industrial metropolis.

How can that sudden takeoff be explained? Cities aren't ideas that spread, viruslike, through larger populations; the town system of the Middle Ages didn't reproduce by spores, the way the city-states of ancient Greece did. And of course, Europe was no longer united by an empire, so there was no command center to decree that a hundred cities should be built in the span of two centuries. How then can we account for the strikingly coordinated urban blossoming of the Middle Ages?

Start by taking the analogies literally. Why does a field of wildflowers suddenly bloom in the spring? Why does water turn to ice? Both systems undergo "phase transitions"—changing from one defined state to another at a critical juncture—in response to changing levels of energy flowing through them. Leave a kettle of water sitting at room temperature in your kitchen, and it will retain its liquid form for weeks. But increase the flow of energy through the kettle by putting it on a hot stove, and within minutes you'll induce a phase transition in the water, transforming it into a gas. Take a field of tall meadow buttercups accustomed to nightly frost and ten hours of sun, then raise the temperature thirty degrees and add four hours of sunlight. After a month or two, your field will be golden yellow with buttercups. A linear increase in energy can produce a nonlinear change in the system that conducts that energy, a change that would be difficult to predict in advance—assuming, that is, you'd never seen a flowering plant before, or a steam room.

The urban explosion of the Middle Ages is an example of the same phenomenon. We saw before that the idea of building cities didn't spread through Europe via word of mouth, but what did spread through Europe, starting around A.D. 1000, were a series of technological advances that combined to produce a dramatic change in the human capacity for harnessing energy flows. As the historian Lynn White Jr. writes, "These innovations . . . consolidated to form a remarkably efficient new way of exploiting the soil." First, the heavy wheeled plow, which tapped the muscular energy of domesticated animals, arrived with the German invaders, then swept through the river valleys north of the Loire; at roughly the same time, European farmers adopted triennial field rotation, which increased land productivity by at least a third. Capturing more energy from the soil meant that larger population densities could be maintained. As larger towns began to form, another soil-based technology became commonplace, one that was even more environmentally friendly: recycling the waste products generated by town residents in the form of crop fertilizer. As Mumford writes, "Wooded areas in Germany, a wilderness in the ninth century, gave way to plowland; the boggy Low Countries, which had supported only a handful of hardy fishermen, were transformed into one of the most productive soils in Europe." The result is a positive feedback loop: the plow and the crop rotation makes better soil, which supplies enough energy to sustain towns, which generate enough fertilizer to make better soil, which generates enough energy to sustain even larger towns.

We sometimes talk about emergent systems "bootstrapping" themselves into existence, but in the case of the Middle Ages, we can safely say that the early village residents shat themselves into full-fledged towns. But those residents aren't setting out to build bigger settlements; they're all solving local problems, such as how to make their fields more productive, or what to do with all the human waste of a busy town. And yet those local decisions combine to form the macrobehavior of the urban explosion. "This acceleration in urban development," writes philosopher-historian Manuel

De Landa, "would not be matched for another five hundred years, when a new intensification of the flow of energy—this time arising from the exploitation of fossil fuels—propelled another great spurt of city birth and growth in the 1800s." And with that new flow of energy, new kinds of cities emerged: the factory towns of Manchester and Leeds, and the great metropolitan superorganisms of London, Paris, and New York.

We are, by all accounts, in the midst of another technological revolution—an information age, a time of near-infinite connectedness. If information storage and retrieval was the latent purpose of the urban explosion of the Middle Ages, it is the manifest purpose of the digital revolution. All of which raises the question, is the Web learning as well? If cities can generate emergent intelligence, a macrobehavior spawned by a million micromotives, what higher-level form is currently taking shape among the routers and fiber-optic lines of the Internet?

I first started thinking about this question a few years ago, during the promotional tour for my last book, *Interface Culture.* As it happened, my book's publisher also specialized in "contemporary spiritual" titles, and so the in-house publicist sent galleys of what I thought was a decidedly un–New Agey book to every New Age radio station, print zine, and ashram in the country. What's more, some of them ended up taking the bait, and so the tour assumed a slightly schizophrenic air: NPR in the morning, followed by a Q&A with alternative magazines like San Francisco's *Magical Blend* in the afternoon.

The questions from the Harmonic Convergence set turned out to be as consistently smart and forward-thinking and technologically adept as any I'd encountered on the rest of the tour. The New Agers were sensitive to the nuances of my argument, and refreshingly indifferent to the latest IPO pricing. (Contrast that with the TV reporters, who seemed incapable of asking me anything other than "What's your take on Yahoo's market cap?") But just when I'd start kicking myself for embarking on the interview with such prejudice, my interlocutors would roll out a Final Question. "You've written a great deal about the Web and its influence on modern society," they'd say. "Do you think, in the long term, that the rise of the Web is leading towards a single, global, holistic consciousness that will unite us all in godhead?" I'd find myself stammering into the microphone, looking for exit signs.

It's a question with only one responsible answer: "I'm not qualified to answer that." And each time I said this, I thought to myself that something was fundamentally flawed about the concept, something close to a category mistake. For there to be a single, global consciousness, the Web itself would have to be getting smarter, and the Web wasn't a single, unified thing—it was just a vast sum of interlinked data. You could debate whether the Web was making us smarter, but that the Web itself might be slouching toward consciousness seemed ludicrous.

But as the years passed, I found that the question kept bouncing around in my head, and slowly I started to warm up to it, in a roundabout way. Some critics, such as Robert Wright, talk about a "global brain" uniting all the world's disparate pools of information, while other visionaries—such as Bill Joy and Ray Kurzweil—believe that the computational powers of digital technology are accelerating at such a rate that large networks of computers may actually become self-aware sometime in the next century.

Did Arthur C. Clarke and *The Matrix* have it right all along? Is the Web itself becoming a giant brain? I still think the answer is no. But now I think it's worth asking why not.

* * *

Begin by jettisoning two habitual ways of thinking about what a brain is. First, forget about gray matter and synapses. When someone like Wright says "giant brain," he means a device for processing and storing information, like the clustered neighborhoods of Florence. Second, accept the premise that brains can be a collective enterprise. Being individual organisms ourselves, we're inclined to think of brains as discrete things, possessed by individual organisms. But both categories turn out to be little more than useful fictions. As we've seen, ants do their "learning" at the colony level—growing less aggressive with age, or rerouting a food assembly line around a disturbance—while the individual ants remain blissfully ignorant of the larger project. The "colony brain" is the sum of thousands and thousands of simple decisions executed by individual ants. The individual ants don't have anything like a personality, but the colonies do.

Replace *ants* with *neurons,* and *pheromones* with *neurotransmitters,* and you might just as well be talking about the human brain. So if neurons can swarm their way into sentient brains, is it so inconceivable that the process might ratchet itself up one more level? Couldn't individual brains connect with one another, this time via the digital language of the Web, and form something greater than the sum of their parts—what the trendy philosopher/priest Teilhard de Chardin called the noosphere? Wright's not exactly convinced that the answer is yes, but he's willing to go on the record that the question is, as he puts it, "noncrazy":

> Today's talk of a giant global brain is cheap. But there's a difference. These days, most people who talk this way are speaking loosely. Tim Berners-Lee, who invented the World Wide Web, has noted parallels between the Web and the structure of the brain, but he insists that "global brain" is mere metaphor. Teilhard de Chardin, in contrast, seems to have been speaking liter-

> ally: humankind was coming to constitute an actual brain—like the one in your head, except bigger. Certainly there are more people today than in Teilhard's day who take the idea of a global brain literally Are they crazy? Was Teilhard crazy? Not as crazy as you might think.

Part of Wright's evidence here is that *Homo sapiens* brains already have a long history of forming higher-level intelligence. Individual human minds have coalesced into "group brains" many times in modern history, most powerfully in the communal gatherings of cities. In Wright's view, the city functions as a kind of smaller-scale trial run for the Web's worldwide extravaganza, like an Andrew Lloyd Webber musical that gets the kinks out in Toronto before opening on Broadway. As in the urban explosion of the Middle Ages, a city is not just an accidental offshoot of growing population density—it's a kind of technological breakthrough in its own right. Sustainable city life ranks high on the list of modern inventions—as world-transforming as the alphabet (which it helped engender) or the Internet (which may well be its undoing). It's no coincidence that the great majority of the last millennium's inventions blossomed in urban settings. Like the folders and file directories of some oversize hard drive, the group brain of city life endowed information with far more structure and durability than it had previously possessed. Wright's position is that the Web looks to be the digital heir to that proud tradition, uniting the world's intellects in a way that would have astonished the early networkers of Florence or Amsterdam. Macrointelligence emerged out of the bottom-up organization of city life, he argues, and it will do the same on the Web.

I'm obviously sympathetic to Wright's argument, but I think it needs clarifying. Emergence isn't some mystical force that comes into being when agents collaborate; as in the freeways vs. sidewalks debate, there are environments that facilitate higher-level intelligence, and environments that suppress it. To the extent that the Web has connected more sentient beings together than any technology before it, you can see it as a kind of global brain. But both brains and cities do more than just connect, because intelligence requires both connectedness *and* organization. Plenty of decentralized systems in the real world spontaneously generate structure as they increase in size: cities organize into neighborhoods or satellites; the neural connections of our brains develop extraordinarily specialized regions. Has the Web followed a comparable path of development over the past few years? Is the Web becoming more organized as it grows?

You need only take a quick look at the NASDAQ most active list to see that the answer is an unequivocal no. The portals and the search engines exist in the first place because the Web is a tremendously disorganized space, a system where the disorder grows right alongside the overall volume. Yahoo and Google function, in a way, as man-made antidotes to the Web's natural chaos—an engineered attempt to restore structure to a system that is incapable of generating structure on its own. This is the oft-noted paradox of the Web: the more information that flows into its reservoirs, the harder it becomes to find any single piece of information in that sea.

Imagine the universe of HTML documents as a kind of city spread out across a vast landscape, with each document representing a building in that space. The Web's city would be more anarchic than any real-world city on the planet—no patches of related shops and businesses; no meatpacking or theater districts; no bohemian communities or upscale brownstones; not even the much-lamented "edge city" clusters of Los Angeles or Tyson's Corner. The Web's city would simply be an undifferentiated mass of data growing more confusing with each new "building" that's erected—so confusing, in fact, that the mapmakers (the Yahoos and Googles of the world) would generate almost as much interest as the city itself.

And if the Web would make a miserable city, it would do even worse as a brain. Here's Steven Pinker, the author of *How the Mind Works,* in a *Slate* dialogue with Wright:

> The Internet is in some ways like a brain, but in important ways not. The brain doesn't just let information ricochet around the skull. It is organized to do something: to move the muscles in ways that allow the whole body to attain the goals set by the emotions. The anatomy of the brain reflects that: it is not a uniform web or net, but has a specific organization in which emotional circuits interconnect with the frontal lobes, which receive information from perceptual systems and send commands to the motor system. This goal-directed organization comes from an important property of organisms you discuss: their cells are in the same reproductive boat, and thus have no "incentive" to act against the interests of the whole body. But the Internet, not being a cohesive replicating system, has no such organization.

Again, the point here is that intelligent systems depend on structure and organization as much as they do on pure connectedness—and that intelligent systems are guided toward particular types of structure by the laws of natural selection. A latter-day Maxwell's Demon who somehow manages to superglue a billion neurons to each other wouldn't build anything like the human brain, because the brain relies on specific clusters to make sense of the world, and those clusters only emerge out of a complex interplay among neurons, the external world, and our genes (not to mention a few thousand other factors). Some systems, such as the Web, are geniuses at making connections but lousy with structure. The technologies behind the Internet—everything from the microprocessors in each Web server to the open-ended protocols that govern the data itself—have been brilliantly engineered to handle dramatic increases in scale, but they are indifferent, if not downright hostile, to the task of creating higher-level order. There is, of course, a neurological equivalent of the Web's ratio of growth to order, but it's nothing you'd want to emulate. It's called a brain tumor.

Still, in the midst of all that networked chaos, a few observers have begun to detect macropatterns in the Web's development, patterns that are invisible to anyone using the Web, and thus mostly useless. The distribution of Web sites and their audiences appears to follow what is called a power law: the top ten most popular sites are ten times larger than the next hundred more popular sites, which are themselves ten times more popular than the next thousand sites. Other online cartographers have detected "hub" and "spoke" patterns in traffic flows. But none of these macroshapes, even if they do exist, actually makes the Web a more navigable or informative system. These patterns may be self-organizing, but they are not *adaptive* in any way. The patterns are closer to a snowflake's intricacy than a brain's neural net: the snowflake self-organizes into miraculously complicated shapes, but it's incapable of becoming a *smarter* snowflake, or a more effective one. It's simply a frozen pattern. Compare that to the living, dynamic patterns of a city neighborhood or the human brain: both shapes have evolved into useful structures because they have been pushed in that direction by the forces of biological or cultural evolution: our brains are masterpieces of emergence because large-brained primates were, on the whole, more likely to reproduce than their smaller-brained competitors; the trade clusters of the modern city proliferated because their inhabitants prospered more than isolated rural craftsmen. There is great power and creative energy in self-organization, to be sure, but it needs to be channeled toward specific forms for it to blossom into something like intelligence.

But the fact that the Web as we know it tends toward chaotic connections over emergent intelligence is not something intrinsic to all computer networks. By tweaking some of the underlying assumptions behind today's Web, you could design an alternative version that could potentially mimic the self-organizing neighborhoods of cities or the differentiated lobes of the human brain—and could definitely reproduce the simpler collective problem-solving of ant colonies. The Web's not inherently disorganized, it's just built that way. Modify its underlying architecture, and the Web might very well be capable of the group-think that Teilhard envisioned.

How could such a change be brought about? Think about Deborah Gordon's harvester ants, or Paul Krugman's model for edge-city growth. In both systems, the interaction between neighbors is two-way: the foraging ant that stumbles across the nest-building ant registers something from the encounter, and vice versa; the new store that opens up next to an existing store influences the behavior of that store, which in turn influences the behavior of the newcomer. Relationships in these systems are mutual: you influence your neighbors, and your neighbors influence you. All emergent systems are built out of this kind of feedback, the two-way connections that foster higher-level learning.

Ironically, it is precisely this feedback that the Web lacks, because HTML-based links are one-directional. You can point to ten other sites from your home page, but there's no way for those pages to know that you're pointing to them, short of you taking the time to fire off an e-mail to their respective webmasters. Every page on the Web contains precise information about the other addresses it points to, and yet, by definition, no page on the Web knows who's pointing back. It's a limitation that would be unimaginable in any of the other systems that we've looked at. It's like a Gap outlet that doesn't realize that J.Crew just moved in across the street, or an ant that remains oblivious to the other ants it stumbles across in its daily wanderings. The intelligence of a harvester ant colony derives

from the densely interconnected feedback between
encounter each other and change their behavior accordi
ordained rules. Without that feedback, they'd be a randc
blage of creatures butting heads and moving on, incapable of displaying the complex behavior that we've come to expect from the social insects. (The neural networks of the brain are also heavily dependent on feedback loops.) Self-organizing systems use feedback to bootstrap themselves into a more orderly structure. And given the Web's feedback-intolerant, one-way linking, there's no way for the network to learn as it grows, which is why it's now so dependent on search engines to rein in its natural chaos.

Is there a way around this limitation? In fact, a solution exists already, although it does nothing to modify the protocols of the Web, but rather ingeniously works around the shortcomings of HTML to create a true learning network that sits on top of the Web, a network that exists on a global scale. Appropriately enough, the first attempt to nurture emergent intelligence online began with the desire to keep the Web from being so forgetful.

You can't really, *truly* understand Brewster Kahle until you've had him show you the server farm in Alexa Internet's basement. Walk down a flight of outdoor steps at the side of an old military personnel-processing building in San Francisco's Presidio, and you'll see an entire universe of data—or at least a bank of dark-toned Linux servers arrayed along a twenty-foot wall. The room itself—moldy concrete, with a few spare windows gazing out at foot level—might have held a lawn mower and some spare file cabinets a few decades ago. Now it houses what may well be the most accurate snapshot of The Collective Intelligence anywhere in the world: thirty terabytes of data, archiving both the Web itself and the patterns of traffic flowing through it.

As the creator of the WAIS (Wide Area Information Server) system, Kahle was already an Internet legend when he launched Alexa in 1996. The Alexa software used collaborative-filtering-like technology to build connections between sites based on user traffic. The results from its technology are showcased in the "related sites" menu option found in most browsers today. Amazon.com acquired Alexa Internet in 1999, but the company remains happily ensconced in its low-tech Presidio offices, World War II temporary structures filled with the smell of the nearby eucalyptus trees. "In just three years we got bigger than the Library of Congress, the biggest library on the planet," Kahle says, arms outstretched in his basement server farm. "So the question is, what do we do now?"

Obsessed with the impermanence of today's datastreams, Kahle (and his partner, Bruce Gilliat) founded Alexa with the idea of taking "snapshots" of the Web and indexing them permanently on huge storage devices for the benefit of future historians. As they developed that project, it occurred to them that they could easily open up that massive database to casual Web surfers, supplementing their Web browsing experience with relevant pages from the archive. Anytime a surfer encountered a "404 Page Not Found" error—meaning that an old page had been deleted or moved—he or she could swiftly consult the Alexa archive and pull up the original page.

To make this possible, Kahle and Gilliat created a small toolbar that launches alongside your Web browser. Once the application detects a URL request, it scurries off to the Alexa servers, where it queries the database for information about the page you're visiting. If the URL request ends in a File Not Found message, the Alexa application trolls through the archives for an earlier version of the page. Kahle dubbed his toolbar a "surf engine"—a tool that accompanies you as you browse—and he quickly realized that he'd stumbled across a program that could do far more than just resuscitate old Web pages. By tracking the surfing patterns of its users, the software could also make connections between Web sites, connections that might otherwise have been invisible, both to the creators of those sites and the people browsing them.

Two months after starting work on Alexa, Kahle added a new button to his toolbar, with the simple but provocative tag "What's Next?" Click on the button while visiting a Marilyn Monroe tribute site, and you'll find a set of links to other Marilyn shrines online; click while you're visiting a community site for cancer survivors, and you'll find a host of other like-minded sites listed in the pull-down menu. How are these connections formed? By watching traffic patterns, and looking for neighbors. The software learns by watching the behavior of Alexa's users: if a hundred users visit *FEED* and then hop over to *Salon,* then the software starts to perceive a connection between the two Web sites, a connection that can be weakened or strengthened as more behavior is tracked. In other words, the associations are not the work of an individual consciousness, but rather the sum total of thousands and thousands of individual decisions, a guide to the Web created by following an unimaginable number of footprints.

It's an intoxicating idea, and strangely fitting. After all, a guide to the entire Web should be more than just a collection of handcrafted ratings. As Kahle says, "Learning from users is the only thing that scales to the size of the Web." And that learning echoes the clustered neighborhoods of Florence or London. Alexa's power of association—this site is *like* these other sites—emerges out of the desultory travels of the Alexa user base; none of those users are deliberately setting out to create clusters of related sites, to endow the Web with much-needed structure. They simply go about their business, and the system itself learns by watching. Like Gordon's harvester ants, the software gets smarter, grows more organized, the more individual surfing histories it tracks. If only a thousand people fire up Alexa alongside their browsers, the recommendations simply won't have enough data behind them to be accurate. But add another ten thousand users to the mix, and the site associations gain resolution dramatically. The system starts to learn.

Let's be clear about what that learning entails, because it differs significantly from the traditional sci-fi portraits of computerized intelligence, both utopian and dystopian. Alexa makes no attempt to simulate human intelligence or consciousness directly. In other words, you don't teach the computer to read or appreciate Web site design. The software simply looks for patterns in numbers, like the foraging ants counting the number of fellow foragers they encounter per hour. In fact, the "intelligence" of Alexa is really the aggregated wisdom of the thousands—or millions—of people who use the system. The computer churns through the millions of ratings in its database, looks for patterns of likes and dislikes, then reports back to the user with its findings.

It's worth noting here that Alexa is not truly a "recommendation agent"; it is not telling you that you'll *like* the five sites that it suggests. It's saying that there's a *relationship* between the site you're currently visiting and the sites listed on the pull-down menu. The clusters that form via Alexa are clusters of association, and the links between them are not unlike the traditional links of hypertext. Think about the semantics of a hypertext link embedded in an online article: when you see that link, you don't translate it as "If you like this sentence, you'll like this page as well." The link isn't recommending another page; it's pointing out that there's a relationship between the sentence you're reading and the page at the other end of the link. It's still up to you to decide if you're interested in the other sites, just as it's up to you to decide which silk merchant you prefer on the Por Santa Maria. Alexa's simply there to show you where the clusters are.

Outside of the video-game world, Alexa may be the most high-profile piece of emergent software to date: the tool was integrated into the Netscape browser shortly after its release, and the company is now applying its technology to the world of consumer goods. But the genre is certainly diversifying. An East Coast start-up called Abuzz, recently acquired by the New York Times digital unit, offers a filtering service that enables people searching for particular information or expertise to track down individuals who might have the knowledge they're looking for. A brilliant site called Everything2 employs a neural-net-like program to create a user-authored encyclopedia, with related entries grouped together, Alexa-style, based on user traffic patterns. Indeed, the Web industry is teeming with start-ups promising to bring like minds together, whether they're searching for entertainment or more utilitarian forms of information. These are the digital-age heirs to the Por Santa Maria.

Old-school humanists, of course, tend to find something alarming in the idea of turning to computers for expert wisdom and cultural sensibility. In most cases, the critics' objections sound like a strangely inverted version of the old morality tales that once warned us against animating machines: Goethe's (and Disney's) sorcerer's apprentice, Hoffmann's sandman, Shelley's Frankenstein. In the contemporary rendition, it's not that the slave technology grows stronger than us and learns to disobey our commands—it's that we deteriorate to the level of the machines. Smart technology makes us dumber.

The critique certainly has its merits, and even among the Net community—if it's still possible to speak of a single Net community—intelligent software remains much villified in some quarters. Decades ago, in a curiously brilliant book, *God and Golem, Inc.,* Norbert Wiener argued that "in poems, in novels, in painting, the brain seems to find itself able to work very well with material that any computer would have to reject as formless." For many people the distinction persists to this day: we look to our computers for number crunching; when we want cultural advice, we're already blessed with plenty of humans to consult. Other critics fear a narrowing of our aesthetic bandwidth, with agents numbly recommending the sites that everyone else is surfing, all the while dressing their recommendations up in the sheep's clothing of custom-fit culture.

But it does seem a little silly to resist the urge to experiment with the current cultural system, where musical taste is usually determined by the marketing departments at Sony and Dreamworks, and expert wisdom comes in the form of Ann Landers columns and the Psychic Hotline. If the computer is, in the end, merely making connections between different cultural sensibilities, sensibilities that were originally developed by humans and not by machines, then surely the emergent software model is preferable to the way most Westerners consume entertainment: by obeying the dictates of advertising. Software like Alexa isn't trying to replicate the all-knowing authoritarianism of Big Brother or HAL, after all—it's trying to replicate the folksy, communal practice of neighbors sharing information on a crowded sidewalk, even if the neighbors at issue are total strangers, communicating to each other over the distributed network of the Web.

The pattern-seeking algorithms of emergent software are already on their way to becoming one of the primary mechanisms in the great Goldberg contraption of modern social life—as familiar to us as more traditional devices like supply and demand, representational democracy, snap polls. Intelligent software already scans the wires for constellations of book lovers or potential mates. In the future, our networks will be caressed by a million invisible hands, seeking patterns in the digital soup, looking for neighbors in a land where everyone is by definition a stranger.

Perhaps this is only fitting. Our brains got to where they are today by bootstrapping out of a primitive form of pattern-matching. As the futurist Ray Kurzweil writes, "Humans are far more skilled at recognizing patterns than in thinking through logical combinations, so we rely on this aptitude for almost all of our mental processes. Indeed, pattern recognition comprises the bulk of our neural circuitry. These faculties make up for the extremely slow speed of human neurons." The human mind is poorly equipped to deal with problems that need to be solved *serially*—one calculation after another—given that neurons require a "reset time" of about five milliseconds, meaning that neurons are capable of only two hundred calculations per second. (A modern PC can do millions of calculations per second, which is why we let them do the heavy lifting for anything that requires math skills.) But unlike most computers, the brain is a massively parallel system, with 100 billion neurons all working away at the same time. That parallelism allows the brain to perform amazing feats of pattern recognition, feats that continue to confound digital computers—such as remembering faces or creating metaphors. Because each individual neuron is so slow, Kurzweil explains, "we don't have time . . . to think too many new thoughts when we are pressed to make a decision. The human brain relies on precomputing its analyses and storing them for future reference. We then use our pattern-recognition capability to recognize a situation as compatible to one we have thought about and then draw upon our previously considered conclusions."

It's conceivable that the software of today lies at the evolutionary foothills of some larger, distributed consciousness to come, like the SKYNET network from the *Terminator* films that "became self-aware on August 15, 1997." Certainly the evidence suggests that genuinely cognizant machines are still on the distant technological horizon, and there's plenty of reason to suspect they may never arrive. But the problem with the debate over machine learning and intelligence is that it has too readily been divided between the mindless software of today and the sentient code of the near future. The Web may never become self-aware in any way that resembles human self-awareness, but that doesn't mean the Web isn't capable of learning. Our networks will grow smarter in the coming years, but smarter in the way that an immune system or a city grows smarter, not the way a child does. That's nothing to apologize for—an adaptive information network capable of complex pattern recognition could prove to be one of the most important inventions in all of human history. Who cares if it never actually learns how to think for itself?

An emergent software program that tracks associations between Web sites or audio CDs doesn't *listen* to music; it follows purchase patterns or listening habits that we supply and lets us deal with the air guitar and the off-key warbling. On some basic human level, that feels like a difference worth preserving. And maybe even one that we won't ever be able to transcend, a hundred years from now or more. But is it truly a difference in kind, or is it just a difference in degree? This is the question that has haunted the artificial intelligence community for decades now, and it hits close to home in any serious discussion of emergent software. Yes, the computer doesn't listen to music or browse the Web; it looks for patterns in data and converts those patterns into information that is useful—or at least aims to be useful—to human beings. Surely this process is miles away from luxuriating in "The Goldberg Variations," or reading *Slate.*

But what is listening to music if not the search for patterns—for harmonic resonance, stereo repetition, octaves, chord progressions—in the otherwise dissonant sound field that surrounds us every day? One tool scans the zeros and ones on a magnetic disc. The other scans the frequency spectrum. What drives each process is a hunger for patterns, equivalencies, likenesses; in each the art emerges out of perceived symmetry. (Bach, our most mathematical composer, understood this better than anyone else.) Will computers ever learn to *appreciate* the patterns they detect? It's too early to tell. But in a world where the information accessible online is doubling every six months, it is clear that some form of pattern-matching—all those software programs scouring the Net for signs of common behavior, relevant ideas, shared sensibilities—will eventually influence much of our mediated lives, maybe even to the extent that the pattern-seekers are no longer completely dependent on the commands of the masters, just as city neighborhoods grow and evolve beyond the direct control of their inhabitants. And where will that leave the software then? What makes music different from noise is that music has patterns, and our ears are trained to detect them. A software application—no matter how intelligent—can't literally hear the sound of all those patterns clicking into place. But does that make its music any less sweet?

4
Listening to Feedback

Late in the afternoon of January 23, 1992, during a campaign stop at the American Brush Company in Claremont, New Hampshire, the ABC political reporter Jim Wooten asked then-candidate Bill Clinton about allegations being made by an ex-cabaret singer named Gennifer Flowers. While rumors of Clinton's womanizing had been rampant among the press corps, Wooten's question was the first time the young Democratic front-runner had been asked about a specific woman. "She claims she had a long-standing affair with you," Wooten said with cameras running. "And she says she tape-recorded the telephone conversations with you in which you told her to deny you had ever had an affair."

Wooten said later that Clinton took the question as though he'd been practicing his answer for months. "Well, first of all, I read the story. It isn't true. She has obviously taken money to change the story, a story she felt so strongly about that she hired a lawyer to protect her good name not very long ago. She did call me. I never initiated any calls to her. . . ." The candidate's denials went on for another five minutes, and then the exchange was over. Clinton had responded to the question, but was it news? Across the country, a furious debate on journalistic ethics erupted: Did unproven allegations about the candidate's sex life constitute legitimate news? And did it matter that the candidate himself had chosen to deny the allegations on camera? A cabaret singer making claims about the governor's adulterous past was clearly tabloid material—but what happened when the governor himself addressed the story?

After two long hours of soul-searching, all three major television networks—along with CNN and PBS's MacNeil/Lehrer show—chose not to mention Wooten's question on their national news broadcast, or to show any of the footage from the exchange. The story had emphatically been silenced by some of the most influential figures in all of mass media. The decision to ignore Gennifer Flowers had been unanimous—even at the network that had originally posed the question. Made ten or twenty years before, a decision of that magnitude could have ended a story in its tracks (assuming the *Washington Post* and the *New York Times* followed suit the next morning). For the story to be revived, it would need new oxygen—some new development that caused it to be reevaluated. Without new news, the Flowers story was dead.

And yet the following day, all three networks opened with Gennifer Flowers as their lead item. Nothing had happened to the story itself: none of the protagonists had revealed any additional information; even Clinton's opponents were surprisingly mute about the controversy. The powers that be in New York and Washington had decided the day before that there was no story—and yet here were Peter Jennings and Tom Brokaw leading their broadcasts with the tale of a former Arkansas beauty queen and her scandalous allegations.

How did such a reversal come to pass? It's tempting to resort to the usual hand-wringing about the media's declining standards, but in this case, the most powerful figures in televised media had at first stuck to the high road. If they had truly suffered from declining standards, the network execs would have put Jim Wooten on the first night. Something pushed them off the high road, and that something was not reducible to a national moral decline or a prurient network executive. Gennifer Flowers rode into the popular consciousness via the *system* of televised news, a system that had come to be wired in a specific way.

What we saw in the winter of 1992 was not unlike watching Nixon sweat his way through the famous televised debate of 1960. As countless critics have observed since, we caught a first glimpse in that exchange of how the new medium would change the substance of politics: television would increase our focus on the interpersonal skills of our politicians and diminish our focus on the issues. With the Flowers affair, though, the medium hadn't changed; the underlying system had. In the late eighties, changes in the flow of information—and particularly the raw footage so essential to televised news—had pushed the previously top-down system toward a more bottom-up, distributed model. We didn't notice until Jim Wooten first posed that question in New Hampshire, but the world of televised news had taken a significant first step toward emergence. In the hierarchical system of old, the network heads could willfully suppress a story if they thought it was best for the American people not to know, but that privilege died with Gennifer Flowers, and not because of lowered standards or sweeps week. It was a casualty of feedback.

It is commonplace by now to talk about the media's disposition toward feeding frenzies, where the coverage of a story naturally begets more coverage, leading to a kind of hall-of-mirrors environment where small incidents or allegations get amplified into Major Events. You can normally spot one of these feedback loops as it nears its denouement, since it almost invariably triggers a surge of self-loathing that washes through the entire commentariat. These self-critical waters seem to rise on something like an annual cycle: think of the debate about the paparazzi and Princess Di's death, or the permanent midnight of "Why Do We Care So Much About O.J.?" But the feedback loops of the 1990s weren't an inevitability; they came out of specific changes in the underlying system of mass media, changes that brought about the first stirrings of emergence—and foreshadowed the genuinely bottom-up systems that have since flourished on the Web. That feedback was central to the process should come as no surprise: all decentralized systems rely extensively on feedback, for both growth and self-regulation.

Consider the neural networks of the human brain. On a cellular level, the brain is a massive network of nerve cells connected by the microscopic passageways of axons and dendrites. A flash of brain activity—thinking of a word, wrestling with a concept, parsing the syntax of the sentence you're reading now—triggers an array of neuronal circuits like traffic routes plotted on the map of the mind. Each new mental activity triggers a new array, and an unimaginably large number of possible neuronal circuits go unrealized over the course of a human life (one reason why the persistent loss of brain cells throughout our adult years isn't such a big deal). But beneath all that apparent diversity, certain circuits repeat themselves again and again. One of the most tantalizing hypotheses in neuroscience today is that the cellular basis of learning lies in the repetition of those circuits. As neurologist Richard Restak explains, "Each thought and behavior is embedded within the circuitry of the neurons, and . . . neuronal activity accompanying or initiating an experience persists in the form of reverberating neuronal circuits, which become more strongly defined with repetition. Thus habit and other forms of memory may consist of the establishment of permanent and semipermanent neuronal circuits." A given circuit may initially be associated with the idea of sandwiches, or the shape of an isosceles triangle—and with enough repetition of that specific circuit, it marks out a fixed space in the brain and thereafter becomes part of our mental vocabulary.

Why do these feedback loops and reverberating circuits happen? They come into being because the neural networks of the brain are densely interconnected: each individual neuron contains links—in the form of axons and synapses—to as many as a thousand other neurons. When a given neuron fires, it relays that charge to all those other cells, which, if certain conditions are met, then in turn relay the charge to their connections, and so on. If each neuron extended a link to one or two fellow neurons, the chance of a reverberating loop would be greatly reduced. But because neurons reach out in so many directions simultaneously, it's far more likely that a given neuron firing will wind its way back to the original source, thus starting the process all over again. The likelihood of a feedback loop correlates directly to the general interconnectedness of the system.

By any measure, the contemporary mediasphere is a densely interconnected system, even if you don't count the linkages of the online world. Connected not just in the sense of so many homes wired for cable and so many rooftops crowned by satellite dishes, but also in the more subtle sense of information being plugged into itself in ever more baroque ways. Since Daniel Boorstin first analyzed the television age in his still-invaluable 1961 work, *The Image,* the world of media journalism has changed in several significant ways, with most of the changes promoting an increase of relays between media outlets. There are far more agents in the system (twenty-four-hour news networks, headline pagers, newsweeklies, Web sites), and far more repackagings and repurposings of source materials, along with an alarming new willingness to relay uncritically other outlets' reporting. Mediated media-critique, unknown in Boorstin's less solipsistic times, and formerly quarantined to early-nineties creations such as CNN's *Reliable Sources* and the occasional Jeff Greenfield segment on *Nightline*, is now regularly the lead story on *Larry King* and *Hardball*. The overall system, in other words, has shifted dramatically in the direction of distributed networks, away from the traditional top-down hierarchies. And the more the media contemplates its own image, the more likely it is that the system will start looping back on itself, like a Stratocaster leaning against the amp it's plugged into.

The upshot of all this is that—in the national news cycle at least—there are no longer any major stories in which the media does not eventually play an essential role, and in many cases the media's knack for self-reflection creates the story itself. You don't need much of an initial impulse to start the whole circuit reverberating. The Gennifer Flowers story is the best example of this process at work. As Tom Rosenstiel reported in a brilliant *New Republic* piece several years ago, the Flowers controversy blossomed because of a shift in the relationship between the national news networks and their local affiliates, a shift that made the entire system significantly more interconnected. Until the late eighties, local news (the six- and eleven-o'clock varieties) relied on the national network for thirty minutes of national news footage, edited according to the august standards of the veterans in New York. Local affiliates could either ignore the national stories or run footage that had been supplied to them, but if the network decided the story wasn't newsworthy, the affiliates couldn't cover it.

All this changed when CNN entered the picture in the mideighties. Since the new network lacked a pool of affiliates to provide breaking news coverage when local events became national stories, Ted Turner embarked on a strategy of wooing local stations with full access to the CNN news feed. Instead of a tightly edited thirty-minute reel, the affiliates would be able to pick and choose from almost anything that CNN cameras had captured, including stories that the executive producers in Atlanta had decided to ignore. The Flowers episode plugged into this newly rewired system, and the results were startling. Local news affiliates nationwide also had access to footage of Clinton's comment, and many of them chose to jump on the story, even as the network honchos in New York and Washington decided to ignore it. "When NBC News political editor Bill Wheatley got home and turned on the eleven P.M. local news that night, he winced: the station NBC owned in New York ran the story the network had chosen not to air the same evening," Rosenstiel writes. "By the next afternoon, even Jim Lehrer of the cautious *MacNeil/Lehrer NewsHour* on PBS told the troops they had to air the Flowers story against their better judgment. 'It's out of my hands,' he said."

The change was almost invisible to Americans watching at home, but its consequences were profound. The mechanism for determining what constituted a legitimate story had been reengineered, shifting from a top-down system with little propensity for feedback, to a kind of journalistic neural net where hundreds of affiliates participated directly in the creation of the story. And what made the circuit particularly vulnerable to reverberation was that the networks themselves mimicked the behavior of the local stations, turning what might have been a passing anomaly into a full-throttle frenzy. That was the moment at which the system began to display emergent behavior. The system began calling the shots, instead of the journalists themselves. Lehrer had it right when he said the Gennifer Flowers affair was "out of my hands." The story was being driven by feedback.

*   *   *

The Flowers affair is a great example of why emergent systems aren't intrinsically good. Tornadoes and hurricanes are feedback-heavy systems too, but that doesn't mean you want to build one in your backyard. Depending on their component parts, and the way they're put together, emergent systems can work toward many different types of goals: some of them admirable, some more destructive. The feedback loops of urban life created the great bulk of the world's most dazzling and revered neighborhoods—but they also have a hand in the self-perpetuating cycles of inner-city misery. Slums can also be emergent phenomena. That's not an excuse to resign ourselves to their existence or to write them off as part of the "natural" order of things. It's reason to figure out a better system. The Flowers affair was an example of early-stage emergence—a system of local agents driving macrobehavior without any central authority calling the shots. But it was not necessarily *adaptive.*

Most of the time, making an emergent system more adaptive entails tinkering with different kinds of feedback. In the Flowers affair, we saw an example of what systems theorists call positive feedback—the sort of self-fueling cycles that cause a note strummed on a guitar to expand into a howling symphony of noise. But most automated control systems rely extensively on "negative feedback" devices. The classic example is the thermostat, which uses negative feedback to solve the problem of controlling the temperature of the air in a room. There are actually two ways to regulate temperature. The first would be to design an apparatus capable of blowing air at various different temperatures; the occupant of the room would simply select the desired conditions and the machine would start blowing air cooled or heated to the desired temperature. The problem with that system is twofold: it requires a heating/cooling apparatus capable of blowing air at precise temperatures, and it is utterly indifferent to the room's existing condition. Dial up seventy-two degrees on the thermostat, and the machine will start pumping seventy-two-degree air into the room—even if the room's ambient temperature is already in the low seventies.

The negative feedback approach, on the other hand, provides a simpler solution, and one that is far more sensitive to a changing environment. (Not surprisingly, it's the technique used by most home thermostats.) Instead of pumping precisely calibrated air into the room, the system works with three states: hot air, cool air, and no air. It takes a reading of the room's temperature, measures that reading against the desired setting, and then adjusts its state accordingly. If the room is colder than the desired setting, the hot air goes on. If it is warmer, the cool air flows out. The system continuously measures the ambient temperature and continuously adjusts its output, until the desired setting has been reached—at which point it switches into the "no air" state, where it remains until the ambient temperature changes for some reason. The system uses negative feedback to home in on the proper conditions—and for that reason it can handle random changes in the environment.

Negative feedback, then, is a way of reaching an equilibrium point despite unpredictable—and changing—external conditions. The "negativity" keeps the system in check, just as "positive feedback" propels other systems onward. A thermostat with no feedback simply pumps seventy-two-degree air into a room, regardless of the room's temperature. An imaginary thermostat driven by positive feedback might evaluate the change in room temperature and follow that lead: if the thermostat noted that the room had grown warmer, it would start pumping hotter air, causing the room to grow even warmer, causing the device to pump hotter air. Next thing you know, the water in the goldfish bowl is boiling. Negative feedback, on the other hand, lets the system find the right balance, even in a changing environment. A cold front comes in, a window is opened, someone lights a fire—any of these things can happen, and yet the temperature remains constant. Instead of amplifying its own signal, the system regulates itself.

We've been wrestling with information as a medium for negative feedback ever since Norbert Wiener published *Cybernetics* in 1949, and Wiener himself had been thinking about the relationship between control and feedback since his war-related research of the early forties. After the Japanese bombed Pearl Harbor and the United States joined the war in earnest, Wiener was asked by the army to figure out a way to train mechanical guns to fire automatically at their targets. The question Wiener found himself answering was this: Given enough information about the target's location and movement, could you translate that data into something a machine could use to shoot a V-2 out of the sky?

The problem was uniquely suited for the adaptability of negative feedback: the targets were a mixture of noise and information, somewhat predictable but also subject to sudden changes. But as it happened, to solve the problem Wiener also needed something that didn't really exist yet: a digital computer capable of crunching the flow of data in real time. With that need in mind, Wiener helped build one of the first modern computers ever created. When the story is told of Wiener's war years, the roots of the modern PC are usually emphasized, for legitimate reasons. But the new understanding of negative feedback that emerged from the ENIAC effort had equally far-reaching consequences, extending far beyond the vacuum tubes and punch cards of early computing.

For negative feedback is not solely a software issue, or a device for your home furnace. It is a way of indirectly pushing a fluid, changeable system toward a goal. It is, in other words, a way of transforming a complex system into a complex *adaptive* system. Negative feedback comes in many shapes and sizes. You can build it into ballistic missiles or circuit boards, neurons or blood vessels. It is, in today's terms, "platform agonistic." At its most schematic, negative feedback entails comparing the current state of a system to the desired state, and pushing the system in a direction that minimizes the difference between the two states. As Wiener puts it near the outset of *Cybernetics*: "When we desire a motion to follow a given pattern, the difference between this pattern and the actually performed motion is used as a new input to cause the part regulated to move in such a way as to bring its motion closer to that given by the pattern." Wiener gave that knack for self-regulation a name: homeostasis.

Your body is a massively complex homeostatic system, using an intricate network of feedback mechanisms to keep itself stable in the midst of dynamically changing situations. Many of those feedback mechanisms are maintained by the brain, which coordinates external stimuli received by sensory organs and responds by triggering appropriate bodily actions. Our sleep cycles, for instance, depend heavily on negative feedback. The body's circadian rhythms—accumulated after millions of years of life on a planet with a twenty-four-hour day—flow out of the central nervous system, triggering regular changes in urine formation, body temperature, cardiac output, oxygen consumption, cell division, and the secretions of endocrine glands. But for some reason, our body clocks are set a little slow: the human circadian cycle is twenty-five hours, and so we rely on the external world to reset our clock every day, both by detecting patterns of light and darkness, and by detecting the more subtle change in the earth's magnetic field, which shifts as the planet rotates. Without that negative feedback pulling our circadian rhythms back into sync, we'd find ourselves sleeping through the day for two weeks out of every month. In other words, without that feedback mechanism, it would be as though the entire human race were permanently trapped in sophomore year of college.

Understanding the body and the mind as a feedback-regulated homeostatic system has naturally encouraged some people to experiment with new forms of artificial feedback. Since the seventies, biofeedback devices have reported changes in adrenaline levels and muscle tension in real time to individuals wired up to special machines. The idea is to allow patients to manage their anxiety or stress level by letting them explore different mental states and instantly see the physiological effects. With a little bit of practice, biofeedback patients can easily "drive" their adrenaline levels up or down just by imagining stressful events, or reaching a meditative state. Our bodies, of course, are constantly adjusting adrenaline levels anyway—the difference with biofeedback is that the conscious mind enters into that feedback process, giving patients more direct control over the levels of the hormone in their system. That can be a means of better managing your body's internal state, but it can also be a process of self-discovery. The one time I tried conventional biofeedback, my adrenaline levels hovered serenely at the middle of the range for the first five minutes of the session; the doctor actually complimented me on having such a normal and well-regulated adrenal system. And then, in the course of our conversation, I made a joke—and instantly my adrenaline levels shot off the charts. At the end of my visit, the therapist handed me a printout of the thirty-minute session, with my changing adrenaline levels plotted as a line graph. It was, for all intents and purposes, a computer graph of my attempts at humor over the preceding half hour: a flat line interrupted by six or seven dramatic spikes, each corresponding to a witticism that I had tossed out to the therapist.

I walked away from the session without having improved myself in any noticeable way, and certainly I hadn't achieved more control over my adrenaline levels. But I'd learned something nonetheless: that without consciously realizing it, I'd already established a simple feedback circuit for myself years ago, when my body had learned that it could give itself a targeted adrenaline rush by making a passing joke in conversation. I thought of all those office meetings or ostensibly serious conversations with friends where I had found myself compulsively making jokes, despite the inappropriate context; I thought of how deeply ingrained that impulse is in my day-to-day personality—and suddenly it seemed closer to a drug addiction than a personality trait, my brain scrambling to put together a cheap laugh to secure another adrenaline fix. In a real sense, our personalities are partially the sum of all these invisible feedback mechanisms; but to begin to understand those mechanisms, you need additional levels of feedback—in this case, a simple line graph plotted by an ordinary PC.

If analyzing indirect data such as adrenaline levels can reveal so much about the mind's ulterior motives, imagine the possibilities of analyzing the brain's activity directly. That's the idea behind the technology of neurofeedback: rather than measure the *results* of the brain's actions, neurofeedback measures brain waves themselves and translates them into computer-generated images and sounds. Certain brain-wave patterns appear in moments of intense concentration; others in states of meditative calm; others in states of distraction, or fear. Neurofeedback—like so many of the systems we've seen—is simply a pattern amplification and recognition device: a series of EEG sensors applied to your skull registers changes in the patterns of your brain waves and transforms them into a medium that you can perceive directly, often in the form of audio tones or colors on a computer screen. As your brain drifts from one state to another, the tone or the image changes, giving you real-time feedback about your brain's EEG activity. With some practice, neurofeedback practitioners can more readily drive their brains toward specific states—because the neurofeedback technology supplies the brain with new data about its own patterns of behavior. Once you've reached a meditative state using neurofeedback, devotees claim, the traditional modes of meditation seem like parallel parking without a rearview mirror—with enough practice, you can pull it off, but you're missing a lot of crucial information.

Were he alive today, I suspect Wiener would be surprised to find that biofeedback and neurofeedback technology are not yet mainstream therapeutic practices. But Wiener also recognized that homeostatis was not exclusively the province of individual human minds and bodies. If systems of neurons could form elaborate feedback mechanisms, why couldn't larger human collectivities? "In connection with the effective amount of communal information," Wiener wrote, "one of the most surprising facts about the body politic is its extreme lack of efficient homeostatic processes." He would have diagnosed the pathology of Gennifer Flowers in a heartbeat. The Flowers episode was an instance of pure positive feedback, unchecked by its opposing force. Each agent's behavior encouraged more like-minded behavior from other agents. There was nothing homeostatic about the process, only the "ever-widening gyre" of positive feedback.

But if positive feedback causes such a ruckus in the media world, how can the brain rely so heavily on the reverberating circuits of neurons? One answer is a familiar term from today's media: *fatigue.* Every neuron in the brain suffers from a kind of regulated impotence: after firing for a stretch, the cell must go through a few milliseconds of inaction, the "absolute refractory period," during which it is immune to outside stimulation. Along with many other ingenious inhibiting schemes that the brain relies on, fatigue is a way of shorting out the reverberating circuit, keeping the brain's feeding frenzy in check.

It is this short circuit that is lacking in the modern media's vast interconnectedness. Stories generate more stories, which generate stories about the coverage of the stories, which generate coverage about the meta-coverage. (Here the brain science seems wonderfully poetic: What better diagnosis for the 24/7 vertigo of media feedback than "lack of fatigue"?) A brain that can't stop reverberating is one way of describing what happens during an epileptic fit; the media version is something like Steven Brill's epic critique of the Lewinsky coverage in the first issue of *Content*: a high-profile media critic launching a new magazine with a high-profile indictment of the media's obsession with its own reporting. If the problem stemmed from errors of judgment made by individual reporters, then a media critique might make sense. But since the problem lies in the media's own tendency for self-amplification, it only makes the problem worse to cover the coverage. It's like firing a pistol in the air to stop a fusillade. Once again, the Flowers affair illustrates the principle: the story wasn't "real news"—according to the network wise men—until other outlets started covering it. The newsworthiness of a given story can't be judged by the play the story is getting on the other channels. Otherwise the gravitational pull of positive feedback becomes too strong, and the loop starts driving the process, more than the reporters *or* the event itself.

It's not overstating things to say that the story that emerged from this loop was a milestone in American history. It's entirely possible that the Flowers controversy would have subsided had Clinton's answer to Jim Wooten been ignored; the Clintons would never have gone on *60 Minutes,* and a whole series of tropes that appeared around the couple (Clinton's philandering, Hillary's anti–Tammy Wynette feminism) might never have found their way into the public mind. Without Gennifer Flowers in Clinton's past, would the Monica Lewinsky affair have played out the same way? Probably not. And if that's the case, then we must ask: What really brought this chain of events about? On the one hand, the answer is simple: individual life choices made by individual people—

Clinton's decision to have an affair, and to break it off, Flowers's decision to go public, Clinton's decision to answer the question—result in a chain of events that eventually stirs up an international news story. But there is another sine qua non here, which is the decision made several years before, somewhere in an office complex in Atlanta, to share the entire CNN news feed with local affiliates. That decision was not quite a "pseudo event," in Boorstin's famous phrase. It was a "system event": a change in the way information flowed through the overall news system. But it was a material change nonetheless.

If you think that Clinton's remarks on Gennifer Flowers should never have been a story, then who are the culprits? Whom do we blame in such a setting? The traditional critiques don't apply here: there's no oak-paneled, cigar-smoke-filled back room where the puppeteers pull their invisible strings; it's not that the television medium is particularly "hot" or "cold"; there was a profit motive behind CNN's decision to share more footage, but we certainly can't write off the Flowers episode as just another tribute to the greed of the network execs. Once again, we return to the fundamental laws of emergence: the behavior of individual agents is less important than the overall system. In earlier times, the channels that connected politicians, journalists, and ordinary citizens were one-way and hierarchical; they lacked the connections to generate true feedback; and too few agents were interacting to create any higher-level order. But the cable explosion of the eighties changed all that. For the first time, the system started to reverberate on its own. The sound was quiet during those initial years and may not have crossed into an audible range until Jim Wooten asked that question. And yet anyone who caught the nightly news on January 24, 1992, picked up its signal loud and clear.

Still, the top-heavy structure of mass media may keep those loops relatively muted for the foreseeable future, at least where the tube is concerned. Feedback, after all, is usually not a television thing. You need the Web to hear it wail.

In June of 1962, a full year after the appearance of *The Death and Life of Great American Cities,* Lewis Mumford published a scathing critique of Jane Jacobs's manifesto in his legendary *New Yorker* column, "The Sky Line." In her prescriptions for a sidewalk-centric urban renewal, "Mother Jacobs"—as Mumford derisively called her—offered a "homemade poultice for the cure of cancer." The *New Yorker* critic had been an early advocate of Jacobs's work, encouraging her to translate her thoughts into a book while she was a junior editor at *Architecture Forum* in the midfifties. But the book she eventually wrote attacked Mumford's much-beloved Ebenezer Howard and his "garden cities," and so Mumford struck back at his onetime protégé with full fury.

At over ten thousand words, Mumford's critique was extensive and wide-ranging, but the central message came down to the potential of metropolitan centers to self-regulate. Jacobs had argued that large cities can achieve a kind of homeostasis through the interactions generated by lively sidewalks; urban planning that attempted to keep people off the streets was effectively destroying the lifeblood of the urban system. Without the open, feedback-heavy connections of street culture, cities quickly became dangerous and anarchic places. Building a city without sidewalks, Jacobs argued, was like building a brain without axons or dendrites. A city without connections was no city at all, at least in the traditional sense of organic city life. Better to build cities that encouraged the feedback loops of sidewalk traffic by shortening the length of blocks and supporting mixed-use zoning.

Mumford was no fan of the housing projects of the postwar era, but he had lost faith in the self-regulatory powers of massive urban systems. Cities with populations in the millions simply put too much stress on the natural homeostatic tendencies of human collectives. In *The City in History*, published around the same period, Mumford had looked back at the Greek city-states, and their penchant for founding new units once the original community reached a certain size—the urban equivalent of reproducing by spores. His attachment to Ebenezer Howard also stemmed from the same lack of confidence in metropolitan self-regulation: the Garden City movement—not entirely unlike the New Urbanist movement of today—was an attempt to provide the energy and dynamism of city life in smaller doses. The Italian hill towns of the Renaissance had achieved an ideal mix of density and diversity while keeping their overall population within reasonable bounds (reined in largely by the walls that surrounded them). These were street-centric spaces with a vibrant public culture, but they remained knowable communities too: small enough to foster a real sense of civic belonging. That kind of organic balance, Mumford argued, was impossible in a city of 5 million people, where the noise and congestion—the sensory overload of it all—drained out the "vitality" from the city streets. "Jacobs forgets that in organisms there is no tissue growth quite as 'vital' or 'dynamic' as cancer growths. . . . The author has forgotten the most essential characteristic of all organic growth—to maintain diversity and balance, the organism must not exceed the norm of its species. Any ecological association eventually reaches the 'climax stage,' beyond which growth without deterioration is not possible."

Like many debates from the annals of urban studies, the Mumford/Jacobs exchange over the "climax stage" of city life mirrors recent developments in the digital realm, as Web-based communities struggle to manage the problems of runaway growth. The first generation of online hangouts—dial-up electronic bulletin boards like ECHO or the Well—were the equivalent of those Italian hill-towns: lively, innovative, contentious places, but also places that remained within a certain practical size. In their heyday before the Web's takeoff, both services hovered around five thousand members, and within that population, community leaders and other public characters naturally emerged: the jokers and the enablers, the fact checkers and the polemicists. These characters—many of them concealed behind playful pseudonyms—served as the equivalent of Jacobs's shopkeepers and bartenders, the regular "eyes on the street" that give the neighborhood its grounding and its familiarity.

These online communities also divided themselves into smaller units organized around specific topics. Like the trade-specific clusters of Savile Row and the Por Santa Maria, these divisions made the overall space more intelligible, and their peculiarities endowed each community with a distinctive flavor. (For the first few years of its existence, the Grateful Dead discussion area on the Well was larger than all the other areas combined.) Because each topic area attracted a smaller subset of the overall population, visiting each one felt like returning to an old block in a familiar part of town, and running into the same cast of characters that you had found there the last time you visited.

ECHO and the Well had a certain homeostatic balance in those early years—powerfully captured in Howard Rheingold's book *The Virtual Community*—and part of that balance came from the community's own powers of self-organization. But neither was a pure example of bottom-up behavior: the topic areas, for instance, were central-planning affairs, created by fiat and not by footprints; both communities benefited from the strong top-down leadership of their founders. That their overall populations never approached a "climax stage" reflected the slow modem-adoption curve of the general public, and the limited marketing budgets at both operations. More important, the elements of each community that did self-regulate had little to do with the underlying software. Anyone

who spent any time on those services in the early nineties wi
tell you that community leaders and other recognizable figures emerged, but that status existed only in the perceptions of the users themselves. The software itself was agnostic when it came to status, but because the software brought hominid minds together—minds that are naturally inclined to establish hierarchies in social relationships—leaders and pariahs began to appear. The software did recognize official moderators for each discussion area, but those too were appointments handed down from above; you applied to the village chieftain for the role that you desired, and if you'd been a productive member of the society, your wish might be granted. Their were plenty of unofficial leaders, to be sure—but where the code was concerned, the only official moderators came straight from the top.

This mix of hierarchy and heterarchy was well suited to ECHO's and the Well's stage of growth. At five thousand members, the community was still small enough to be managed partially from above, and small enough that groups and recognizable characters naturally emerged. At that scale, you didn't need to solve the problem of self-regulation with software tools: all you needed was software that connected people's thoughts—via the asynchronous posts of a threaded discussion board—and the community could find its own balance. If something went wrong, you could always look to the official leaders for guidance. But even in those heady early days of the virtual community, the collective systems of ECHO and the Well fell short of achieving real homeostasis, for reasons that would become endemic to the next generation of communities then forming on the Web itself.

A threaded discussion board turns out to be an ideal ecosystem for that peculiar species known as the crank—the ideologue obsessed with a certain issue or interpretive model, who has no qualms about interjecting his or her worldview into any discussion,

no day job or family life to keep him from posting
mentary at the slightest provocation. We all know
, the ones grinding their ax from the back of the
r the coffee shop: the conspiracy theorist, the rabid
libertarian, the evangelist—the ones who insist on bringing all conversations back to their particular issue, objecting to any conversation that doesn't play by their rules. In real life, we've developed a series of social conventions that keep the crank from dominating our conversations. For the most pathological cases, they simply don't get invited out to dinner very often. But for the borderline case, a subtle but powerful mechanism is at work in any face-to-face group conversation: if an individual is holding a conversation hostage with an irrelevant obsession, groups can naturally establish a consensus—using words, body language, facial expressions, even a show of hands—making it clear that the majority of the group feels their time is being wasted. The face-to-face world is populated by countless impromptu polls that take the group's collective pulse. Most of them happen so quickly that we don't even know that we're participating in them, and that transparency is one reason why they're as powerful as they are. In the face-to-face world, we are all social thermostats: reading the group temperature and adjusting our behavior accordingly.

Some of those self-regulatory social skills translate into cyberspace—particularly in a threaded discussion forum or an e-mail exchange, where participants have the time and space to express their ideas in long form, rather than in the spontaneous eruptions of real-time chat. But there is a crucial difference in an environment like ECHO or the Well—or in the discussion areas we built at *FEED*. In a public discussion thread, not all the participants are visible. A given conversation may have five or six active contributors and several dozen "lurkers" who read through the posts but don't chime in with their own words. This creates a fundamental imbalance in the system of threaded discussion and gives the crank an opportunity to dominate the space in a way that would be much more difficult off-line. In a threaded discussion, you're speaking both to the other active participants and to the lurkers, and however much you might offend or bore your direct interlocutors, you can always appeal to that silent majority out there—an audience that is both present and absent at the same time. The crank can cling to the possibility that everyone else tuning in is enthralled by his prose, while the active participants can't turn to the room and say, "Show of hands: Is this guy a lunatic or what?"

The crank exploits a crucial disparity in the flow of information: while we conventionally think of threaded discussions as two-way systems, for the lurkers that flow follows a one-way path. They hear us talking, but we hear nothing of them: no laughs, no hisses, no restless stirring, no snores, no rolling eyeballs. When you factor in the lurkers, a threaded discussion turns out to be less interactive than a traditional face-to-face lecture, and significantly less so than a conversation around a dinner table, where even the most reticent participants contribute with gestures and facial expressions. Group conversations in the real world have an uncanny aptitude for reaching a certain kind of homeostasis: the conversation moves toward a zone that pleases as much of the group as possible and drowns out voices that offend. A group conversation is a kind of circuit board, with primary inputs coming from the official speakers, and secondary inputs coming from the responses of the audience and other speakers. The primary inputs adjust their signal based on the secondary inputs of group feedback. Human beings—for reasons that we will explore in the final section—are exceptionally talented at assessing the mental states of other people, both through the direct exchanges of spoken language and the more oblique feedback mechanisms of gesture and intonation. That two-way exchange gives our face-to-face group conversations precisely the flexibility and responsiveness that Wiener found lacking in mass communications.

I suspect Wiener would immediately have understood the virtual community's problem with cranks and lurkers. Where the Flowers affair was a case of runaway positive feedback, the tyranny of the crank results from a scarcity of feedback: a system where the information flows are unidirectional, where the audience is present and at the same time invisible. These liabilities run parallel to the problems of one-way linking that we saw in the previous chapter. Hypertext links and virtual communities were supposed to be the advance guard of the interactive revolution, but in a real sense they only got halfway to the promised land. (Needless to say, the ants were there millions of years ago.) And if the cranks and obsessive-compulsives flourish in a small-scale online community of several thousand members, imagine the anarchy and noise generated by a million community members. Surely there is a "climax stage" on that scale where the online growth turns cancerous, where the knowable community becomes a nightmare of overdevelopment. If feedback couldn't help regulate the digital villages of early online communication, what hope can it possibly have on the vast grid of the World Wide Web?

The sleepy college town of Holland, Michigan, might seem like the last place you'd expect to generate a solution for the problem of digital sprawl, but the Web has never played by the rules of traditional geography. Until recent years, Holland had been best known for its annual tulip festival. But it is increasingly recognized as the birthplace of Slashdot.org—the closest thing to a genuinely self-organizing community that the Web has yet produced.

Begun as a modest bulletin board by a lifetime Hollander named Rob Malda, Slashdot came into the world as the ultimate in knowable communities: just Malda and his friends, discussing programming news, Star Wars rumors, video games, and other geek-chic marginalia. "In the beginning, Slashdot was small," Malda writes. "We got dozens of posts each day, and it was good. The signal was high, the noise was low." Before long, though, Slashdot floated across the rising tsunami of Linux and the Open Source movement and found itself awash in thousands of daily visitors. In its early days, Slashdot had felt like the hill towns of ECHO and the Well, with strong leadership coming from Malda himself, who went by the handle Commander Taco. But the volume of posts became too much for any single person to filter out the useless information. "Trolling and spamming became more common," Malda says now, "and there wasn't enough time for me to personally keep them in check and still handle my other responsibilities."

Malda's first inclination was to create a Slashdot elite: twenty-five handpicked spam warriors who would sift through the material generated by the community, eliminating irrelevant or obnoxious posts. While the idea of an elite belonged to a more hierarchical tradition, Malda endowed his lieutenants with a crucial resource: they could rate other contributions, on a scale of -1 to 5. You could browse through Slashdot.org with a "quality filter" on, effectively telling the software, "Show me only items that have a rating higher than 3." This gave his lieutenants a positive function as well as a negative one. They could emphasize the good stuff and reward users who were productive members of the community.

Soon, though, Slashdot grew too large for even the elites to manage, and Malda went back to the drawing board. It was the kind of thing that could only have happened on the Web. A twenty-two-year-old college senior, living with a couple of buddies in a low-rent house—affectionately dubbed Geek House One—in a nondescript Michigan town, creates an intimate online space for his friends to discuss their shared obsessions, and within a year fifty thousand people each day are angling for a piece of the action. Without anything resembling a genuine business infrastructure, much less a real office, Malda needed far more than his twenty-five lieutenants to keep the Slashdot community from descending into complete anarchy. But without the resources to hire a hundred full-time moderators, Slashdot appeared to be stuck at the same impasse that Mumford had described thirty years before: stay small and preserve the quality of the original community; keep growing and sacrifice everything that had made the community interesting in the first place. Slashdot had reached its "climax stage."

What did the Commander do? Instead of expanding his pool of special authorized lieutenants, he made *everyone* a potential lieutenant. He handed over the quality-control job to the community itself. His goals were relatively simple, as outlined in the Frequently Asked Questions document on the site:

1. Promote quality, discourage crap.
2. Make Slashdot as readable as possible for as many people as possible.
3. Do not require a huge amount of time from any single moderator.
4. Do not allow a single moderator a "reign of terror."

Together, these objectives define the parameters of Slashdot's ideal state. The question for Malda was how to build a homeostatic system that would naturally push the site toward that state without any single individual being in control. The solution that he arrived at should be immediately recognizable by now: a mix of negative and positive feedback, structured randomness, neighbor interactions, and decentralized control. From a certain angle, Slashdot today resembles an ant colony. From another, it looks like a virtual democracy. Malda himself likens it to jury duty.

Here's how it works: If you've spent more than a few sessions as a registered Slashdot user, the system may on occasion alert you that you have been given moderator status (not unlike a jury summons arriving in your mailbox). As in the legal analogy, moderators only serve for a finite stretch of time, and during that stretch they have the power to rate contributions made by other users, on a scale of -1 to 5. But that power diminishes with use: each moderator is endowed only with a finite number of points that he or she can distribute by rating user contributions. Dole out all your ratings, and your tenure as a moderator comes to an end.

Those ratings coalesce into something that Malda called karma: if your contributions as a user are highly rated by the moderators, you earn karma in the system, giving you special privileges. Your subsequent posts begin life at a higher rating than usual, and you are more likely to be chosen as a moderator in future sessions. This last privilege exemplifies meta-feedback at work, the ratings snake devouring its own tail: moderators rate posts, and those ratings are used to select future moderators. Malda's system not only encouraged quality in the submissions to the site; it also set up an environment where community leaders could naturally rise to the surface. That elevation was specifically encoded in the software. Accumulating karma on Slashdot was not just a metaphor for winning the implicit trust of the Slashdot community; it was a quantifiable number. Karma had found a home in the database.

Malda's point system brings to mind the hit points of Dungeons & Dragons and other classics of the role-playing genre. (That the Slashdot crowd was already heavily versed in the role-playing idiom no doubt contributed a great deal to the rating system's quick adoption.) But Malda had done something more ambitious than simply porting gaming conventions to the community space. He had created a kind of currency, a pricing system for online civics. By ensuring that the points would translate into special privileges, he gave them value. By making one's moderation powers expendable, he created the crucial property of scarcity. With only one or the other, the currency is valueless; combine the two, and you have a standard for pricing community participation that actually works.

The connection between pricing and feedback is itself more than a metaphor. As a character in Jane Jacobs's recent Socratic dialogue, *The Nature of Economies,* observes: "Adam Smith, back in 1775, identified prices of goods and rates of wages as feedback information, although of course he didn't call it that because the word *feedback* was not in the vocabulary at the time. But he understood the idea. . . . In his sober way, Smith was clearly excited about the marvelous form of order he'd discovered, as well he should have been. He was far ahead of naturalists in grasping the principle of negative feedback controls."

Malda himself claims that neither *The Wealth of Nations* nor *The Dungeon Master's Guide* were heavy in his thoughts in Geek House One. "There wasn't really anything specific that inspired me," Malda says now. "It was mostly trial and error. The real influence was my desire to please users with very different expectations for Slashdot. Some wanted it to be Usenet: anything goes and unruly. Others were busy people who only wanted to read three to four comments a day." You can see the intelligence and flexibility of the system firsthand: visit the Slashdot site and choose to view all the posts for a given conversation. If the conversation is more than a few hours old, you'll probably find several hundred entries, with at least half of them the work of cranks and spammers. Such is the fate of any Web site lucky enough to attract thousands of posts an hour.

Set your quality threshold to four or five, however, and something miraculous occurs. The overall volume drops precipitously—sometimes by an order of magnitude—but the dozen or two posts that remain will be as stimulating as anything you've read on a traditional content site where the writers and the editors are actually paid to put their words and arguments together. It's a miracle not so much because the quality is lurking there somewhere in the endless flood of posting. Rather, it's a miracle because the community has collectively done such an exceptional job at bringing that quality to light. In the digital world, at least, there is life after the climax stage.

Slashdot is only the beginning. In the past two years, user ratings have become the kudzu of the Web, draping themselves across pages everywhere you look. Amazon had long included user ratings for all the items in its inventory, but in 1999 it began to let users rate the reviews of other users. An ingenious site called Epinions cultivates product reviews from its audience and grants "trust" points to contributors who earn the community's respect. The online auction system of eBay utilizes two distinct feedback mechanisms layered on top of each other: the price feedback of the auction bids coupled to the user ratings that evaluate buyers and sellers. One system tracks the value of stuff; the other tracks the value of people.

Indeed, the adoption rate for these feedback devices is accelerating so rapidly that I suspect in a matter of years a Web page without a dynamic rating system attached will trigger the same response that a Web page without hyperlinks triggers today: yes, it's technically possible to create a page without these features, but what's the point? The Slashdot system might seem a little complex, a little esoteric for consumers who didn't grow up playing D&D, but think of the millions of people who learned how to use a computer for the first time in the past few years, just to get e-mail or to surf the Web. Compared to that learning curve, figuring out the rules of Slashdot is a walk in the park.

And rules they are. You can't think of a system like the one Malda built at Slashdot as a purely *representational* entity, the way you think about a book or a movie. It is partly representational, of course: you read messages via the Slashdot platform, and so the components of the textual medium that Marshall McLuhan so brilliantly documented in *The Gutenberg Galaxy* are on display at Slashdot as well. Because you are reading words, your reception of the information behind those words differs from what it would have been had that information been conveyed via television. The medium is still the message on Slashdot—it's just that there's another level to the experience, a level that our critical vocabularies are only now finding words for.

In a Slashdot-style system, there is a medium, a message, and an audience. So far, no different from television. The difference is that those elements exist alongside a set of rules that govern the way the messages flow through the system. "Interactivity" doesn't do justice to the significance of this shift. A button that lets you e-mail a response to a published author; a tool that lets you build your own home page; even a collection of interlinked pages that let you follow your own path through them—these are all examples of interactivity, but they're in a different category from the self-organizing systems of eBay or Slashdot. Links and home-page-building tools are cool, no question. But they are closer to a newspaper letters-to-the-editor page than Slashdot's collective intelligence.

First-generation interactivity may have given the consumer a voice, but systems like Slashdot force us to accept a more radical proposition: to understand how these new media experiences work, you have to analyze the message, the medium, *and the rules.* Think of those thousand-post geek-Dionysian frenzies transformed into an informative, concise briefing via the Slashdot quality filters. What's interesting here is not just the medium, but rather the rules that govern what gets selected and what doesn't. It's an algorithmic problem, then, and not a representational one. It is the difference between playing a game of Monopoly and hanging a Monopoly board on your wall. There are representational forces unleashed by a game of Monopoly (you have to be able to make out the color coding of the various properties and to count your money) but what makes the game interesting—indeed, what makes it a game at all—lies in the instruction set that you follow while playing. Slashdot's rules are what make the medium interesting—so interesting, in fact, that you can't help thinking they need their own category, beyond message *and* medium.

Generically, you can describe those rules as a mix of positive and negative feedback pushing the system toward a particular state based on the activities of the participants. But the mix is different every time. The edge cities of Paul Krugman's model used feedback to create polycentric clusters, while other metropolitan systems collapse into a single, dense urban core. The networks in CNN-era television have engendered runaway positive feedback loops such as the Gennifer Flowers story, while a system like Slashdot achieves homeostatic balance, at least when viewed at level 5. Different feedback systems produce different results—even when those systems share the same underlying medium. In the future, every Web site may well be connected to a rating mechanism, but that doesn't mean all Web sites will behave the same way. There may be homeostasis at Slashdot's level 5, but you can always choose to read the unfiltered, anarchic version at level -1.

Is there a danger in moving to a world where all our media responds directly to user feedback? Some critics, such as *The Control Revolution*'s Andrew Shapiro, worry about the tyranny of excessive user personalization, as in the old Nicholas Negroponte vision of the *Daily Me,* the newspaper perfectly custom-tailored to your interests—so custom-tailored, in fact, that you lose the serendipity and surprise that we've come to expect from reading the newspaper. There's no stumbling across a different point of view, or happening upon an interesting new field you knew nothing about—the *Daily*

Me simply feeds back what you've instructed the software to find, and nothing more. It's a mind-narrowing experience, not a mind-expanding one. That level of personalization may well be around the corner, and we'll take a closer look at its implications in the conclusion. But for now, it's worth pointing out that the Slashdot system is indifferent to your personal interests—other than your interest in a general level of quality. The "ideal state" that the Slashdot system homes in on is not defined by an individual's perspective; it is defined by the overall group's perspective. The collective decides what's quality and what's crap, to use Rob Malda's language. You can tweak the quality-to-crap ratio based on your individual predilections, but the ratings themselves emerge through the actions of the community at large. It's more groupthink than *Daily Me.*

Perhaps, then, the danger lies in too much groupthink. Malda designed his system to evaluate submissions based on the average Slashdot reader—although the karma points tend to select moderators who have a higher-than-average reputation within the community. It's entirely possible that Malda's rules have created a tyranny of the majority at Slashdot, at least when viewed at level 5. Posts that resonate with the "average" Slashdotter are more likely to rise to the top, while posts that express a minority viewpoint may be demoted in the system. (Technically, the moderation guidelines suggest that users should rate posts based purely on quality, not on whether they agree with the posts, but the line is invariably a slippery one.) From this angle, then, Slashdot bears a surprising resemblance to the old top-down universe of pre-cable network television. Both systems have a heavy center that pulls content toward the interests of the "average user"—like a planet pulling satellites into its orbit. In the days before cable fragmentation, the big three networks were competing for the entire television-owning audience, which encouraged them to serve up programming designed for the average viewer rather than for a particular niche. (McLuhan observed how this phenomenon was pushing the political parties toward the center as well.) The network decision to pursue the center rather than the peripheries was invariably made at the executive level, of course—unlike at Slashdot, where the centrism comes from below. But if you're worried about suppressing diversity, it doesn't really matter whether it comes from above or below. The results are the same, either way. Majority viewpoints get amplified, while minority viewpoints get silenced.

This critique showcases why we need a third term beyond *medium* and *message.* While it's true that Slashdot's filtering software creates a heavy center, that tendency is not inherent to the Web medium, or even the subset of online communities. You could just as easily build a system that would promote both quality *and* diversity, simply by tweaking the algorithm that selects moderators. Change a single variable in the mix, and a dramatically different system emerges. Instead of picking moderators based on the average rating of their posts, the new system picks moderators whose contributions have triggered the greatest range of responses. In this system, a member who was consistently rated highly by the community would be unlikely to be chosen as a moderator, while a member who inspired strong responses either way—both positive and negative—would be first in line to moderate. The system would reward controversial voices rather than popular ones. You'd still have moderators deleting useless spam and flamebait, and so the quality filters would remain in place. But the fringe voices in the community would have a stronger presence at level 5, because the feedback system would be rewarding perspectives that deviate from the mainstream, that don't aim to please everyone all the time. The cranks would still be marginalized, assuming their polemics annoyed almost everyone who came across them. But the thoughtful minorities—the ones who attract both admirers *and* detractors—would have a place at the table.

There's no reason why centrist Slashdot and diverse Slashdot can't coexist. If you can adjust the quality filters on the fly, you could just as easily adjust the diversity filters. You could design the system to track the ratings of both popular and controversial moderators; users would then be able to view Slashdot through the lens of the "average" user on one day, and through the lens of a more diverse audience the next. The medium and the message remain the same; only the rules change from one system to the other. Adjust the feedback loops, and a new type of community appears on the screen. One setting gives you Gennifer Flowers and cyclone-style feeding frenzies, another gives you the shapeless datasmog of Usenet. One setting gives you an orderly, centrist community strong on shared values, another gives you a multiculturalist's fantasy. As Wiener recognized a half century ago, feedback systems come in all shapes and sizes. When we come across a system that doesn't work well, there's no point in denouncing the use of feedback itself. Better to figure out the specific rules of the system at hand and start thinking of ways to wire it so that the feedback routines promote the values we want promoted. It's the old sixties slogan transposed into the digital age: if you don't like the way things work today, change the system.

5

Control Artist

On the screen, the pixels dance: bright red dots with faint trails of green, scurrying across a black background, like fireflies set against the sky of a summer night. For a few seconds, the movement on-screen looks utterly random: pixels darting back and forth, colliding, and moving on. And then suddenly a small pocket of red dots gather together in a pulsing, erratic circle, ringed by a strip of green. The circle grows as more red pixels collide with it; the green belt expands. Seconds later, another lopsided circle appears in the corner of the screen, followed by three more. The circles are unlike any geometric shape you've ever seen. They seem more like a life-form—a digital blob—pulsing haphazardly, swelling and contracting. Two blobs slowly creep toward each other, then merge, forming a single unit. After a few minutes, seven large blobs dominate, with only a few remaining free-floating red pixels ambling across the screen.

Welcome to the world of Mitch Resnick's tool for visualizing self-organizing systems, StarLogo. A descendant of Seymour Papert's legendary turtle-based programming language, Logo, StarLogo allows you to model emergent behavior using simple, English-like commands—and it displays that behavior in vivid, real-time animations. If decentralized systems can sometimes seem counterintuitive or abstract, difficult to describe in words, StarLogo makes them come to life with dynamic graphics that are uniquely suited for the Nintendo generation. If a calendar is a tool for helping us think about the flow of time, and a pie chart is a tool for thinking about statistical distributions, StarLogo is a tool for thinking about bottom-up systems. And, in fact, those lifelike blobs on the screen take us back to the very beginnings of our story: they are digital slime molds, cells aggregating into larger clusters without any "pacemaker" cell leading the way.

"Those red pixels are the individual slime mold cells," Resnick says, pointing at the screen, sitting in his Cambridge office. "They're programmed to wander aimlessly around the screen space, and as they wander, they 'emit' the green color, which quickly fades away. That color is the equivalent of the c-AMP chemical that the molds use to coordinate their behavior. I've programmed the red cells to 'sniff' the green color and follow the gradient in the color. 'Smelling' the green pixels leads the cells toward each other."

Like Gordon's ant colonies, Resnick's slime mold simulation is sensitive to population density. "Let's start with only a hundred slime mold cells," he says, adjusting a slider on the screen that alters the number of cells in the simulation. He presses a start button, and a hundred red pixels begin their frenetic dance—only this time, no clusters appear. There are momentary flashes of green as a few cells collide, but no larger shapes emerge at all.

"With a hundred cells, there isn't enough contact for the aggregates to form. But triple the population like so," he says, pulling the slider farther to the right, "and you increase the contact between cells. At three hundred cells, you'll usually get one cluster after a few minutes, and sometimes two." We wait for thirty seconds or so, and after a few false starts, a cluster takes shape near the center of the screen. "Once they come together, the slime molds are extremely difficult to break apart, even though they can be very fickle about aggregating in the first place."

Resnick then triples the population and starts the simulation over again. It's a completely different system this time around: there's a flash of red-celled activity, then almost immediately ten clusters form, nearly filling the screen with pulsing watermelon shapes. Only a handful of lonely red cells remain, drifting aimlessly between the clusters. More is *very* different. "The interesting thing is," Resnick says with a chuckle, "you wouldn't have necessarily predicted that behavior in advance, just from looking at the instructions. You might have said, the slime mold cells will all immediately form a giant cluster, or they'll form clusters that keep breaking up. In fact, neither is the case, and the whole system turns out to be much more sensitive to initial conditions. At a hundred cells, there are no clusters at all; at three hundred, you'll probably get one, but it'll be pretty much permanent; and at nine hundred cells, you'll immediately get ten clusters, but they'll bounce around a little more." But you couldn't tell any of that just by looking at the original instruction set. You have to make it *live* before you can understand how it works.

StarLogo may look like a video game at first glance, but Resnick's work is really more in the tradition of Friedrich Froebel, the German educator who invented kindergarten, and who spent much of his career in the early nineteenth century devising ingenious toys that would both amuse and entertain toddlers. "When Froebel designed the first kindergarten," Resnick tells me, "he developed a set of toys they called Froebel's gifts, and he carefully designed them with the assumption that the object he'd put in the hands of kids would make a big difference in what they learned and how they learned. We see the same thing carried through today. We see some of our new technology as the latter-day versions of Froebel's gifts, trying to put new sorts of materials and new types of toys in the hands of kids that will change what they think about—and the *way* they think about it."

StarLogo, of course, is designed to help kids—and grown-ups, for that matter—think about a specific type of phenomenon, but it is by no means limited to slime molds. There are StarLogo programs that simulate ant foraging, forest fires, epidemics, traffic jams—even programs that generate more traditional Euclidean shapes using bottom-up techniques. (Resnick calls this "turtle geometry," after the nickname used to describe the individual agents in a StarLogo program, a term that is itself borrowed from the original Logo language, which Papert designed to teach children about traditional programming techniques.) This knack for shape-shifting is one of the language's great virtues. "StarLogo is a type of modeling environment where kids can build models of certain phenomena that they might observe in the world," Resnick says. "Specifically, it enables them to build models of phenomena where lots of things interact with each other. So they might model cars on a highway, or they might model something like a bird flock, where the kids design behavior for lots of individual birds and then see the patterns that form through all the interactions.

"One reason that we're especially interested in building a tool like this is that these phenomena are common in the everyday world," he continues. "We see bird flocks and traffic jams all of the time. On the other hand, people have a lot of trouble understanding these types of phenomena. When people see a flock of birds, they assume the bird in the front is the leader and the others are just following. But that's not the way the real birds form flocks. In fact, each bird just follows simple rules and they end up together as a group."

At its core, StarLogo is optimized for modeling emergent systems like the ones we've seen in the previous chapters, and so the building blocks for any StarLogo program are familiar ones: local interactions between large numbers of agents, governed by simple rules of mutual feedback. StarLogo is a kind of thinking prosthetic, a tool that lets the mind wrap itself around a concept that it's not naturally equipped to grasp. We need StarLogo to help us understand emergent behavior for the same reason we need X-ray machines or calculators: our perceptual and cognitive faculties can't do the work on their own.

It's a limitation that can be surprisingly hard to overcome. Consider the story that Resnick tells of artificial-intelligence guru Marvin Minsky encountering the slime mold simulation for the first time. "One day shortly after I developed the first working prototype of StarLogo, Minsky wandered into my office. On the computer screen he saw an early version of my StarLogo slime mold program. There were several green blobs on the screen (representing a chemical pheromone), with a cluster of turtles moving around inside each blob. A few turtles wandered randomly in the empty space between the blobs. Whenever one of these turtles wandered close enough to a blob, he joined the cluster of turtles inside."

Minsky scanned the screen for a few seconds, then asked Resnick what he was working on. "I explained that I was experimenting with some self-organizing systems. Minsky looked at the screen for a while, then said, 'But those creatures aren't self-organizing. They're just moving toward the green food.'"

"Minsky had assumed that the green blobs were pieces of food, placed throughout the turtles' world. In fact, the green blobs were created by the turtles themselves. But Minsky didn't see it that way. Instead of seeing creatures organizing themselves, he saw the creatures organized around some preexisting pieces of food. He assumed that the pattern of aggregation was determined by the placement of food. And he stuck with that interpretation—at least for a while—even after I told him that the program involved self-organization."

Minsky had fallen for the myth of the ant queen: the assumption that collective behavior implied some kind of centralized authority—in this case, that the food was dictating the behavior of the slime mold cells. Minsky assumed that you could predict where the clusters would form by looking at where the food was placed when the simulation began. But there wasn't any food. Nor was there anything dictating that clusters should form in specific locations. The slime mold cells were self-organizing, albeit within parameters that Resnick had initially defined.

"Minsky has thought more—and more deeply—about self-organization and decentralized systems than almost anyone else," Resnick writes. "When I explained the rules underlying the slime mold program to him, he understood immediately what was happening. But his initial assumption was revealing. The fact that even *Marvin Minsky* had this reaction is an indication of the powerful attraction of centralized explanations."

Of course, on the most fundamental level, StarLogo is itself a centralized system: it obeys rules laid down by a single authority—the programmer. But the route from Resnick's code to those slime mold clusters is indirect. You don't program the slime mold cells to form clusters; you program them to follow patterns in the trails left behind by their neighbors. If you have enough cells, and if the trails last long enough, you'll get clusters, but they're not something you can control directly. And predicting the number of clusters—or their longevity—is almost impossible without extensive trial-and-error experimentation with the system. Kevin Kelly called his groundbreaking book on decentralized behavior *Out of Control*, but the phrase doesn't quite do justice to emergent systems—or at least the ones that we've deliberately set out to create on the computer screen. Systems like StarLogo are not utter anarchies: they obey rules that we define in advance, but those rules only govern the micromotives. The macrobehavior is another matter. You don't control that directly. All you can do is set up the conditions that you think will make that behavior possible. Then you press play and see what happens.

That kind of oblique control is a funny thing to encounter in the world of software, but it is becoming increasingly common. Programming used to be thought of as a domain of pure control: you told the computer what to do, and the computer had no choice but to obey your orders. If the computer failed to do your bidding, it inevitably had to do with a bug in your code, and not the machine's autonomy. The best programmers were the ones who had the most control of the system, the ones who could compel the machines to do the work with the least amount of code. It's no accident that Norbert Wiener derived the term *cybernetics* from the Greek word for "steersman": the art of software has from the beginning been about control systems and how best to drive them.

But that control paradigm is slowly giving way to a more oblique form of programming: software that you "grow" instead of engineer, software that learns to solve problems autonomously, the way Oliver Selfridge envisioned with his Pandemonium model. The new paradigm borrows heavily from the playbook of natural selection, breeding new programs out of a varied gene pool. The first few decades of software were essentially creationist in philosophy—an almighty power wills the program into being. But the next generation is profoundly Darwinian.

* * *

he program for number sorting devised several years ago ›mputing legend Danny Hillis, a program that under- of our conventional assumptions about how software should be produced. For years, number sorting has served as one of the benchmark tests for ingenious programmers, like chess-playing applications. Throw a hundred random numbers at a program and see how many steps it takes to sort the digits into the correct order. Using traditional programming techniques, the record for number sorting stood at sixty steps when Hillis decided to try his hand. But Hillis didn't just sit down to write a number-sorting application. What Hillis created was a recipe for learning, a program for creating another program. In other words, he didn't teach the computer how to sort numbers. He taught the computer to figure out how to sort numbers *on its own.*

Hillis pulled off this sleight of hand by connecting the formidable powers of natural selection to a massively parallel supercomputer—the Connection Machine that he himself had helped design. Instead of authoring a number-sorting program himself—writing out lines of code and debugging—Hillis instructed the computer to generate thousands of miniprograms, each composed of random combinations of instructions, creating a kind of digital gene pool. Each program was confronted with a disorderly sequence of numbers, and each tried its hand at putting them in the correct order. The first batch of programs were, as you might imagine, utterly inept at number sorting. (In fact, the overwhelming majority of the programs were good for nothing at all.) But some programs were better than others, and because Hillis had established a quantifiable goal for the experiment—numbers arranged in the correct order—the computer could select the few programs that were in the ballpark. Those programs became the basis for the next iteration, only Hillis would also mutate *their* code slightly and crossbreed them with the other promising programs. And then the whole process would repeat itself: the most successful programs of the new generation would be chosen, then subjected to the same transformations. Mix, mutate, evaluate, repeat.

After only a few minutes—and thousands of cycles—this evolutionary process resulted in a powerful number-sorting program, capable of arranging a string of random numbers in seventy-five steps. Not a record breaker, by any means, but impressive nonetheless. The problem, though, was that the digital gene pool was maxing out at the seventy-five-step mark. Each time Hillis ran the sequence, the computer would quickly evolve a powerful and efficient number sorter, but it would run out of steam at around seventy-five steps. After enough experimentation, Hillis recognized that his system had encountered a hurdle often discussed by evolutionary theorists: the software had stumbled across a local maximum in the fitness landscape.

Imagine the space of all possible number-sorting programs spread out like a physical landscape, with more successful programs residing at higher elevations, and less successful programs lurking in the valleys. Evolutionary software is a way of blindly probing that space, looking for gradients that lead to higher elevations. Think of an early stage in Hillis's cycle: one evolved routine sorts a few steps faster than its "parent" and so it survives into the next round. That survival is the equivalent of climbing up one notch on the fitness landscape. If its "descendant" sorts even more efficiently, its "genes" are passed on to the next generation, and it climbs another notch higher.

The problem with this approach is that there are false peaks in the fitness landscape. There are countless ways to program a computer to sort numbers with tolerable efficiency, but only a few ways to sort numbers if you're intent on setting a world record. And those different programs vary dramatically in the way they tackle the problem. Think of those different approaches as peaks on the fitness landscape: there are thousands of small ridges, but only a few isolated Everests. If a program evolves using one approach, its descendants may never find their way to another approach—because Hillis's system only rewarded generations that *improved* on the work done by their ancestors. Once the software climbs all the way to the top of a ridge, there's no reward in descending and looking for another, higher peak, because a less successful program—one that drops down a notch on the fitness landscape—would instantly be eliminated from the gene pool. Hillis's software was settling in at the seventy-five-step ridges because the penalty for searching out the higher points was too severe.

Hillis's stroke of genius was to force his miniprograms out of the ridges by introducing predators into the mix. Just as in real-world ecosystems, predators effectively raised the bar for evolved programs that became lazy because of their success. Before the introduction of predators, a miniprogram that had reached a seventy-five-step ridge knew that its offspring had a chance of surviving if it stayed at that local maximum, but faced almost certain death if it descended to search out higher ground. But the predators changed all that. They hunted down ridge dwellers and forced them to improvise: if a miniprogram settled into the seventy-five-step range, it could be destroyed by predator programs. Once the predators appeared on the scene, it became more productive to descend to lower altitudes to search out a new peak than to stay put at a local maximum.

Hillis structured the predator-prey relationship as an arms race: the higher the sorting programs climbed, the more challenging the predators became. If the system stumbled across a seventy-step peak, then predators were introduced that hunted down seventy-step programs. Anytime the software climbers decided to rest on their laurels, a predator appeared to scatter them off to find higher elevations.

After only thirty minutes of this new system, the computer had evolved a batch of programs that could sort numbers in sixty-two steps, just two shy of the all-time record. Hillis's system functioned, in biological terms, more like an environment than an organism: it created a space where intelligent programs could grow and adapt, exceeding the capacities of all but the most brilliant flesh-and-blood programmers. "One of the interesting things about the sorting programs that evolved in my experiment is that I do not understand how they work," Hillis writes in his book *The Pattern on the Stone*. "I have carefully examined their instruction sequences, but I do not understand them: I have no simpler explanation of how the programs work than the instruction sequences themselves. It may be that the programs are not understandable."

Proponents of emergent software have made some ambitious claims for their field, including scenarios where a kind of digital Darwinism leads to a simulated intelligence, capable of open-ended learning and complex interaction with the outside world. (Most advocates don't think that such an intelligence will necessarily resemble *human* smarts, but that's another matter, one that we'll examine in the conclusion.) In the short term, though, emergent software promises to transform the way that we think about creating code: in the next decade, we may well see a shift from top-down, designed programs to bottom-up, evolved versions, like Hillis's number-sorting applet—"less like engineering a machine," Hillis says, "than baking a cake, or growing a garden."

That transformation may be revolutionary for the programmers, but if it does its job, it won't necessarily make much of a difference for the end users. We might notice our spreadsheets recalculating a little faster and our grammar checker finally working, but we'll be dealing with the end results of emergent software, not the process itself. (The organisms, in Darwinian terms, and not the environment that nurtured them.) But will ordinary computer-users get a chance to experiment with emergent software firsthand, a chance to experiment with its more oblique control systems? Will growing gardens of code ever become a mainstream activity?

In fact, we can get our hands dirty already. And we can do it just by playing a game.

It's probably fair to say that digital media has been wrestling with "control issues" from its very origins. The question of control, after all, lies at the heart of the interactive revolution, since making something interactive entails a shift in control, from the technology—or the puppeteers behind the technology—to the user. Most recurring issues in interactive design hover above the same underlying question: Who's driving here, human or machine? Programmer or user? These may seem like esoteric questions, but they have implications that extend far beyond design-theory seminars or cybercafé philosophizing. I suspect that we're only now beginning to understand how complicated these issues are, as we acclimate to the strange indirection of emergent software.

In a way, we've been getting our sea legs for this new environment for the past few years now. Some of the most interesting interactive art and games of the late nineties explicitly challenged our sense of control or made us work to establish it. Some of these designs belonged to the world of avant-garde or academic experimentation, while others had more mainstream appeal. But in all these designs, the feeling of wrestling with or exploring the possibilities of the software—the process of mastering the system—was transformed from a kind of prelude to the core experience of the design. It went from a bug to a feature.

There are different ways to go about challenging our sense of control. Some programs, such as the ingenious Tap, Type, Write—created by MIT's John Maeda—make it immediately clear that the user is driving. The screen starts off with an array of letters; hitting a specific key triggers a sudden shift in the letterforms presented on-screen. The overall effect is like a fireworks show sponsored by Alphabet Soup. Press a key, and the screen explodes, ripples, reorders itself. It's hypnotic, but also a little mystifying. What algorithm governs this interaction? Something happens on-screen when you type, but it takes a while to figure out what rules of transformation are at work here. You know you're doing something, you just don't know what it is.

The OSS code, created by the European avant-punk group Jodi.org, messes with our sense of control on a more profound—some would say annoying—level. A mix of anarchic screen-test patterns and eclectic viral programming, the Jodi software is best described as the digital equivalent of an aneurysm. Download the software and the desktop overflows with meaningless digits; launch one of the applications, and your screen descends instantly into an unstable mix of static and structure. Move the mouse in one direction, or double click, and there's a fleeting sense of something changing. Did the flicker rate shift? Did those interlaced patterns reverse themselves? At hard-to-predict moments, the whole picture show shuts down—invariably after a few frantic keystrokes and command clicks—and you're left wondering, Did I do that?

No doubt many users are put off by the dislocations of Tap, Type, Write and OSS, and many walk away from the programs feeling as though they never got them to work quite right, precisely because their sense of control remained so elusive. For me, I find these programs strangely empowering; they challenge the mind in the same way distortion challenged the ear thirty-five years ago when the Beatles and the Velvet Underground first began overloading their amps. We find ourselves reaching around the noise—the lack of structure—for some sort of clarity, only to realize that it's the reaching that makes the noise redemptive. Video games remind us that messing with our control expectations can be fun, even addictive, as long as the audience has recognized that the confusion is part of the show. For a generation raised on MTV's degraded images, that recognition comes easily. The Nintendo generation, in other words, has been well prepared for the mediated control of emergent software.

Take as example one of the most successful titles from the Nintendo64 platform, Shigeru Miyamoto's Zelda: Ocarina of Time. Zelda embodies the uneven development of late-nineties interactive entertainment. The plot belongs squarely to the archaic world of fairy tales—a young boy armed with magic spells sets off to rescue the princess. As a control system, though, Zelda is an incredibly complex structure, with hundreds of interrelated goals and puzzles dispersed throughout the game's massive virtual world. Moving your character around is simple enough, but figuring out what you're supposed to do with him takes hours of exploration and trial and error. By traditional usability standards, Zelda is a complete mess: you need a hundred-page guidebook just to establish what the rules are. But if you see that opacity as part of the art—like John Cale's distorted viola—then the whole experience changes: you're exploring the world of the game and the rules of the game at the same time.

Think about the ten-year-olds who willingly immerse themselves in Zelda's world. For them, the struggle for mastery over the system doesn't feel like a struggle. They've been decoding the landscape on the screen—guessing at causal relations between actions and results, building working hypotheses about the system's underlying rules—since before they learned how to read. The conventional wisdom about these kids is that they're more nimble at puzzle solving and more manually dexterous than the TV generation, and while there's certainly some truth to that, I think we lose something important in stressing how talented this generation is with their joysticks. I think they have developed another skill, one that almost looks like patience: they are more tolerant of being out of control, more tolerant of that exploratory phase where the rules don't all make sense, and where few goals have been clearly defined. In other words, they are uniquely equipped to embrace the more oblique control system of emergent software. The hard work of tomorrow's interactive design will be exploring the tolerance—that suspension of control—in ways that enlighten us, in ways that move beyond the insulting residue of princesses and magic spells.

With these new types of games, a new type of game designer has arisen as well. The first generation of video games may have indirectly influenced a generation of artists, and a handful were adopted as genuine objets d'art, albeit in a distinctly campy fashion. (Tabletop Ms. Pac-Man games started to appear at downtown Manhattan clubs in the early nineties, around the time the Museum of the Moving Image created its permanent game collection.) But artists themselves rarely ventured directly into the game-design industry. Games were for kids, after all. No self-respecting artist would immerse himself in that world with a straight face.

But all this has changed in recent years, and a new kind of hybrid has appeared—a fusion of artist, programmer, and complexity theorist—creating interactive projects that challenge the mind and the thumb at the same time. And while Tap, Type, Write and Zelda were not, strictly speaking, emergent systems, the new generation of game designers and artists have begun explicitly describing their work using the language of self-organization. This too brings to mind the historical trajectory of the rock music genre. For the first fifteen or twenty years, the charts are dominated by lowest-common-denominator titles, rarely venturing far from the established conventions or addressing issues that would be beyond the reach of a thirteen-year-old. And then a few mainstream acts begin to push at the edges—the Beatles or the Stones in the music world, Miyamoto and Peter Molyneux in the gaming community—and the expectations about what constitutes a pop song or a video game start to change. And that transformation catches the attention of the avant-garde—the Velvet Underground, say, or the emergent-game designers—who suddenly start thinking of pop music or video games as a legitimate channel for self-expression. Instead of writing beat poetry or staging art happenings, they pick up a guitar—or a joystick.

By this standard, Eric Zimmerman is the Lou Reed of the new gaming culture. A stocky thirty-year-old, with short, club-kid hair and oversize Buddy Holly glasses, Zimmerman has carved out a career for himself that would have been unthinkable even a decade ago: bouncing between academia (he teaches at NYU's influential Interactive Telecommunications Program), the international art scene (he's done installations for museums in Geneva, Amsterdam, and New York), and the video-game world. Unlike John Maeda and or Jodi.org, Zimmerman doesn't "reference" the iconography of gaming in his work—he openly embraces that tradition, to the extent that you have to think of Zimmerman's projects as games first and art second. They can be fiendishly fun to play and usually involve spirited competition between players. But they are also self-consciously designed as emergent systems.

"One of the pleasures of what I do," Zimmerman tells me, over coffee near the NYU campus, "is that you get to see a player take what you've designed and use it in completely unexpected ways." The designer, in other words, controls the micromotives of the player's actions. But the way those micromotives are exploited—and the macrobehavior that they generate—are out of the designer's control. They have a life of their own.

Take Zimmerman's game Gearheads, which he designed during a brief sojourn at Phillips Interactive in 1996. Gearheads is a purebred emergent system: a meshwork of autonomous agents following simple rules and mutually influencing each other's behavior. It is a close relative of StarLogo or Gordon's harvester ants, but it's ingeniously dressed up to look like a modern video game. Instead of spare colored pixels, Zimmerman populated the Gearhead world with an eclectic assortment of children's toys that march across the screen like a motley band of animated soldiers.

"There are twelve windup toys," Zimmerman explains. "You design a box of toys by choosing four of them. You wind up your toy and release it from the edges of the game board, and the goal of the game is to get as many toys as possible across your opponent's side of the screen. Each of the toys has a unique set of behaviors that affect the behavior of other toys." A skull toy, for instance, "frightens" toys that it encounters, causing them to reverse direction, while an animated hand winds up other toys, allowing them to march across the screen for a longer duration. As with the harvester ants or the slime mold cells, when one agent encounters another agent, both agents may launch into a new pattern of behavior. Stumble across your hundredth forager of the afternoon, and you'll switch over to midden duty; stumble across Zimmerman's skull toy and you'll turn around and go the other way.

"The key thing is that once you've released your toys, they're autonomous. You're only affecting the system from the margins," Zimmerman says. "It's a little chaos machine: unexpected things happen, and you only control it from the edges." As Zimmerman tested Gearheads in early 1996, he found that this oblique control system resulted in behavior that Zimmerman hadn't deliberately programmed, behavior that emerged out of the local interactions of the toys, despite the overall simplicity of the game.

"Two toys reverse the direction of other toys—the skull, and the Santa toy, who's called Krush Kringle," Zimmerman says. "He walks for a few steps and then he pounds the ground, and all the toys near him reverse direction. During our testing, we found a combination where you could release one Krush Kringle out there, then the walking hand that winds up toys, then another Krush Kringle. The hand would run out and wind up the first Krush, and then the Krush would pound the floor, reversing the direction of the hand, and sending it back to the second Krush, which it would wind up. Then the second Krush would stomp on the ground, and the hand would turn around and wind up the first Krush. And so the little system of these three toys would march together across the screen, like a small flock of birds. The first time we saw it happen, we were astonished."

These unexpected behaviors may not seem like much at first glance, particularly in a climate that places so much emphasis on photo-realistic, 3-D worlds and blood-spattering combat. Zimmerman's toys are kept deliberately simple; they don't simulate intelligence, and they don't trigger symphonies of surround sound through your computer speakers. A snapshot of Resnick's slime molds looks like something you might have seen on a first-generation Atari console. But I'll put my money on the slime molds and Krush Kringles nonetheless. Those watermelon clusters and autowinding flocks strike me as the very beginning of what will someday form an enormously powerful cultural lineage. Watching these patterns emerge spontaneously on the screen is a little like watching two single-celled organisms decide to share resources for the first time. It doesn't look like much, but the same logic carried through a thousand generations, or a hundred thousand—like Hillis growing his gardens of code—can end up changing the world. You just have to think about it on the right scale.

Most game players, alas, live on something close to day-trader time, at least when they're in the middle of a game—thinking more about their next move than their next meal, and usually blissfully oblivious to the ten- or twenty-year trajectory of software development. No one wants to play with a toy that's going to be fun after a few decades of tinkering—the toys have to be engaging *now,* or kids will find other toys. And one of the things that make all games so engaging to us is that they have rules. In traditional games like Monopoly or go or chess, the fun of the game—the play—is what happens when you explore the space of possibilities defined by the rules. Without rules, you have something closer to pure improv theater, where anything can happen at any time. Rules give games their structure, and without that structure, there's no game: every move is a checkmate, and every toss of the dice lands you on Park Place.

This emphasis on rules might seem like the antithesis of the open-ended, organic systems we've examined over the preceding chapters, but nothing could be further from the truth. Emergent systems too are rule-governed systems: their capacity for learning and growth and experimentation derives from their adherence to low-level rules: ants choosing to forage or not, based on patterns in their encounters with other ants; the Alexa software making connections based on patterns in the clickstream. If any of these systems—or, to put it more precisely, the agents that make up these systems—suddenly started following their own rules, or doing away with rules altogether, the system would stop working: there'd be no global intelligence, just a teeming anarchy of isolated agents, a swarm without logic. Emergent behaviors, like games, are all about living within the boundaries defined by rules, but also using that space to create something greater than the sum of its parts.

Understanding emergence should be a great boon for the videogame industry. But some serious challenges face the designers of games that attempt to harness the power and adaptability of self-organization and channel it into a game aimed at a mass audience. And those challenges all revolve around the same phenomenon: the capacity of emergent systems to suddenly start behaving in unpredictable ways, sorcerer's-apprentice style—like Zimmerman's flock of Krush Kringles.

Consider the case of Evolva, a widely hyped game released in mid-2000 by a British software company called Computer Artworks. The product stood as something of a change for CA, which was last seen marketing a trippy screen-saver called Organic Art that allowed you to replace your desktop with a menagerie of alien-looking life-forms. That program came bundled with a set of prepackaged images, but more adventurous users could also grow their own, "breeding" new creatures with the company's A-Life technology. While the Organic Art series was a success, it quickly became clear to the CA team that *interacting* with your creatures would be much more entertaining than simply gazing at snapshots of them. Who wants to look at Polaroids of Sea-Monkeys when you can play with the adorable little critters yourself?

And so Computer Artworks turned itself into a video-game company. Evolva was their first fully interactive product to draw upon the original artificial-life software, integrating its mutation and interbreeding routines into a game world that might otherwise be mistaken for a hybrid of Myth and Quake. The plot was standard-issue video-game fare: Earth has been invaded by an alien parasite that threatens world destruction; as a last defense, the humans send out packs of fearless "genohunters" to save the planet. Users control teams of genohunters, occupying the point of view of one while issuing commands to the others. A product of biological engineering themselves, genohunters are capable of analyzing the DNA of any creature they kill and absorbing useful strands into their own genetic code. Once you've absorbed enough DNA, you can pop over to the "mutation" screen and tinker with your genetic makeup—adding new genes and mutating your existing ones, expanding your character's skills in the process. It's like suddenly learning how to program in C++, only you have to eat the guy from tech support to see the benefits.

That appetite for DNA gives the A-Life software its entrée into the gameplay. "As the player advances through the game, new genes are collected and added to the available gene pool," lead programmer Rik Heywood explained to me in an e-mail conversation. "When the player wants to modify one of their creations, they can go to the mutation screen. Starting from the current set of DNA, two new generations can be created by combining the DNA from the existing genohunter with the DNA in the collected gene pool and some slight random mutations. The new sets of DNA are used to morph the skin, grow appendages all over the body, and develop new abilities, such as breathing fire or running faster."

The promotional material for Evolva makes a great deal of noise about this open-endedness. Some 14 billion distinct characters can be generated using the mutation screen, which means that unless Computer Artists strikes a licensing deal with other galaxies, players who venture several levels deep in the game will be playing with genetically unique genohunters. For the most part, those mutations result in relatively superficial external changes, more like a new paint job than an engine overhaul. The more sophisticated alterations to the genohunters' behavior—fire-breathing, laser-shooting, long-distance jumping, among others—are largely discrete skills programmed directly by the CA team. You won't see any genohunters spontaneously learning how to play the cello or use sonar. The bodies of your genohunters may end up looking dramatically different from where they started, but those bodies won't let their hosts adopt radically new skills.

These limitations may well make the game more enjoyable. For a sixteen-year-old Quake player who's just trying to kill as many parasites as possible on his way to the next level, suddenly learning how to read braille is only going to be a distraction. Anyone who has spent time playing a puzzle-based narrative game like Myst knows nothing is more frustrating than spending two hours trying to solve a puzzle that you don't yet have the tools to solve, because you haven't stumbled across them in your explorations of the game space. Imagine how much more frustrating to get stumped by a puzzle because you haven't evolved gills or lock-picking skills yet. In a purely open-ended system—where the tools may or may not evolve depending on the whims of natural selection—that frustration would quickly override any gee-whiz appeal of growing your own characters. And so Heywood and his team have planted DNA for complex skills near puzzles or hurdles that require those skills. "For example, if we wanted to be sure that the player had developed the ability to breath fire by a particular point in the game," he explains, "we would block the path with some flammable plants and place some creatures with a fire-breathing ability nearby."

The blind watchmaker of Evolva's mutation engine turns out to have some sight after all. Heywood's solution might be the smartest short-term move for the gamers, but it's worth pointing out that it also runs headlong against the principles of Darwinism. Not only are you playing God by deliberately selecting certain traits over others, but the DNA for those traits is planted near the appropriate obstacles. It's like some strange twist on Lamarckian evolution: the giraffe neck grows longer each generation, but only because the genes for longer necks happen to sprout next to the banana trees. The space of possibility unleashed by an open-ended Darwinian engine was simply too large for the rule-space of the game itself. A game where anything can happen is by definition not a game.

Is there a way to reconcile the unpredictable creativity of emergence with the directed flow of gaming? The answer, I think, will turn out to be a resounding yes, but it's going to take some trial and error.

One way involves focusing on traditional emergent systems—such as flocks and clusters—and less on the more open-ended landscape of natural selection. Evolva is actually a great example of the virtues of this sort of approach. Behind the scenes, each creature in the Evolva world is endowed with sensory inputs and emotive states: fear, pain, aggression, and so on. Creatures also possess memories that link those feelings with other characters, places, or actions—and they are capable of sharing those associations with their comrades. As the web of associations becomes more complex, and more interconnected, new patterns of collective behavior can evolve, creating a lifelike range of potential interactions between creatures in the world.

"Say you encounter a lone creature," Heywood explains. "When you first meet it, it is maybe feeling very aggressive and runs in to attack your team. However, you have it outnumbered and start causing it some serious pain. Eventually fear will become the dominant emotion, causing the creature to run away. It runs around a corner and meets a large group of friends. It communicates with these other creatures, informing them of the last place it saw you. Being in a large group of friends brings its fear back down, and the whole group launches a new attack on the player." The *group* behavior can evolve in unpredictable ways, based on external events and each creature's emotional state, even if the virtual DNA of those creatures remains unchanged. There is something strangely comforting in this image, particularly for anyone who thinks social patterns influence our behavior as readily as our genes do. Heywood had to restrict the artificial-life engine because the powers of natural selection are too unpredictable for the rules-governed universe of a video game. But building an emergent system to simulate collective behavior among characters actually improved the gameplay, made it more lifelike without making it impossible. Emergence trumps "descent with modification": you may not be able to use Evolva's mutation engine to grow wings, but your creatures can still learn new ways to flock.

There is a more radical solution to this problem, though, and it's most evident in the god-games genre. Classic games like SimCity—or 1999's best-selling semi-sequel The Sims, which lets game players interact with simulated personalities living in a small neighborhood—have dealt with the unpredictability of emergent software by eliminating predefined objectives altogether. You define your own goals in these games; you're not likely to get stuck on a level because you haven't figured out how to "grow" a certain resource, for the simple reason that there are no preordained levels to follow. You define your own hurdles as you play. In SimCity, you decide whether to build a megalopolis or a farming community; whether to build an environmentally correct new urbanist village or a digital Coketown. Of course, you may find it hard to achieve those goals as you build the city, but because those goals aren't part of the game's official rules, you don't feel stuck in the same way that you might feel stuck in Evolva, staring across the canyon without the genes for jumping.

There's a catch here, though. "The challenge is, the more autonomous the system, the more autonomous the virtual creatures, the more irrelevant the player is," Zimmerman explains. "The problem with a lot of the 'god games' is that it's difficult to feel like you're having a meaningful impact on the system. It's like you're wearing these big, fuzzy gloves and you're trying to manipulate these tiny little objects." Although it can be magical to watch a Will Wright simulation take on a life of its own, it can also be uniquely frustrating—when that one neighborhood can't seem to shake off its crime problem, or your Sims refuse to fall in love. For better or worse, we control these games from the edges. The task of the game designer is to determine just how far off the edge the player should be.

Nowhere is this principle more apparent than in the control

panel that Will Wright built for The Sims. Roll your cursor along the bottom of the screen while surveying your virtual neighborhood, and a status window appears, with the latest info on your characters' emotional and physical needs: you'll see in an instant whether they've showered today, or whether they're pining for some companionship. A click away from that status window is a control panel screen, where you can adjust various game attributes. A "settings" screen is by now a standard accoutrement of any off-the-shelf game: you visit the screen to adjust the sound quality or the graphics resolution, or to switch difficulty levels. At first glance, the control panel for The Sims looks like any of these other settings screens: there's a button that changes whether the window scrolls automatically as you move the mouse, and another that turns off the background music. But alongside these prosaic options, there is a toggle switch that says, in unabashed Cartesian terms, "Free will."

If you leave "Free will" off, The Sims quickly disintegrates into a nightmare of round-the-clock maintenance, requiring the kind of constant attention you'd expect in a nursery or a home for Alzheimer's patients. Without free will, your Sims simply sit around, waiting for you to tell them what to do. They may be starving, but unless you direct them to the fridge, they'll just sit out their craving for food like a gang of suburban hunger artists. Even the neatest of the Sims will tolerate piles of rotting garbage until you specifically order them to take out the trash. Without a helpful push toward the toilet, they'll even relieve themselves right in the middle of the living room.

Playing The Sims without free will selected is a great reminder that too much control can be a disastrous thing. But the opposite can be even worse. Early in the design of The Sims, Wright recognized that his virtual people would need a certain amount of autonomy for the game to be fun, and so he and his team began developing a set of artificial-intelligence routines that would allow the Sims to think for themselves. That AI became the basis for the character's "free will," but after a year of work, the designers found that they'd been a little too successful in bringing the Sims to life.

"One of our biggest problems here was that our AI was too smart," Wright says now. "The characters chose whichever action would maximize their happiness at any given moment. The problem is that they're usually much better at this than the player." The fun of The Sims comes from the incomplete information that you have about the overall system: you don't know exactly what combination of actions will lead to a maximum amount of happiness for your characters—but the software behind the AI can easily make those calculations, because the happiness quota is built out of the game's rules. In Wright's early incarnations of the game, once you turned on free will, your characters would go about maximizing their happiness in perfectly rational ways. The effect was not unlike hiring Deep Blue to play a game of chess for you—the results were undeniably good ones, but where was the fun?

And so Wright had to dumb down his digital creations. "We did it in two ways," he says. "First, we made them focus on immediate gratification rather than long-term goals—they'd rather sit in front of the TV and be couch potatoes than study for a job promotion. Second, we gave their personality a very heavy weight on their decisions, to an almost pathological degree. A very neat Sim will spend way too much time picking up—even after other Sims—while a sloppy Sim will never do this. These two things were enough to ensure that the player was a sorely needed component—ambition? balance?—of their world." In other words, Wright made their decisions local ones and made the rules that governed their behavior more intransigent. For the emergent system of the game to work, Wright had to make the Sims more like ants than people.

I think there is something profound, and embryonic, in that "free will" button, and in Wright's battle with the autonomy of his creations—something both like and unlike the traditional talents that we expect from our great storytellers. Narrative has always been about the mix of invention and repetition; stories seem like stories because they follow rules that we've learned to recognize, but the stories that we most love are ones that surprise us in some way, that break rules in the telling. They are a mix of the familiar and the strange: too much of the former, and they seem stale, formulaic; too much of the latter, and they cease to be stories. We love narrative genres—detective, romance, action-adventure—but the word *generic* is almost always used as a pejorative.

It misses the point to think of what Will Wright does as storytelling—it doesn't do justice to the novelty of the form, and its own peculiar charms. But that battle over control that underlies any work of emergent software, particularly a work that aims to entertain us, runs parallel to the clash between repetition and invention in the art of the storyteller. A good yarn surprises us, but not too much. A game like The Sims gives its on-screen creatures some autonomy, but not too much. Emergent systems are not stories, and in many ways they live by very different rules, for both creator and consumer. (For one, emergent systems make that distinction a lot blurrier.) But the art of the storyteller can be enlightening in this context, because we already accept the premise that storytelling *is* an art, and we have a mature vocabulary to describe the gifts of its practitioners. We are only just now developing such a language to describe the art of emergence. But here's a start: great designers like Wright or Resnick or Zimmerman are *control* artists—they have a feel for that middle ground between free will and the nursing home, for the thin line between too much order and too little. They have a feel for the edges.

PART THREE

Screenshot from SimCity 3000 *(Courtesy of Maxis)*

Can a selectional system be simulated? The answer must be split into two parts. If I take a particular animal that is the result of evolutionary and developmental selection, so that I already know its structure and the principles governing its selective processes, I can simulate the animal's structure in a computer. But a system undergoing selection has two parts: the animal or organ, and the environment or world. No instructions come from events of the world. No instructions come from events of the world to the system on which selection occurs. Moreover, events occurring in an environment or a world are unpredictable. How then do we simulate events and their effects on selection? One way is as follows:

1. Simulate the organ or the animal as described above, making provision for the fact that, as a selective system, it contains a generator of diversity—mutations, alterations in neural wiring, or synaptic changes that are unpredictable.
2. Independently simulate a world or environment constrained by known physical principles, but allow for the occurrence of unpredictable events.
3. Let the simulated organ or animal interact with the simulated world or the real world without prior information transfer, so that selection can take place.
4. See what happens.

—Gerald Edelman

6

The Mind Readers

What are you thinking about right now? Because my words are being communicated to you via the one-way medium of the printed page, this is a difficult question for me to answer. But if I were presenting this argument while sitting across a table from you, I'd already have an answer, or at least an educated guess—even if you'd been silent the entire time. Your facial gestures, eye movements, body language, would all be sending a steady stream of information about your internal state—signals that I would intuitively pick up and interpret. I'd see your eyelids droop during the more contorted arguments, note the chuckle at one of my attempts at humor, register the way you sit upright in the chair when my words get your attention. I could no more prohibit my mind from making those assessments than you could stop your mind from interpreting my spoken words as language. (Assuming you're an English speaker, of course.) We are both locked in a communicational dance of ɔth—and yet, amazingly, we're barely aware of the

ːs are innate mind readers. Our skill at imagining ːntal states ranks up there with our knack for language and our opposable thumbs. It comes so naturally to us and has engendered so many corollary effects that it's hard for us to think of it as a special skill at all. And yet most animals lack the mind-reading skills of a four-year-old child. We come into the world with a genetic aptitude for building "theories of other minds," and adjusting those theories on the fly, in response to various forms of social feedback.

In the mideighties, the UK psychologists Simon Baron-Cohen, Alan Leslie, and Uta Frith conducted a landmark experiment to test the mind-reading skills of young children. They concealed a set of pencils within a box of Smarties, the British candy. They asked a series of four-year-olds to open the box and make the unhappy discovery of the pencils within. The researchers then closed the box up and ushered a grown-up into the room. The children were then asked what the grown-up was expecting to find within the Smarties box—not what they *would* find, mind you, but what they were *expecting* to find. Across the board, the four-year-olds gave the right answer: the clueless grown-up was expecting to find Smarties, not pencils. The children were able to separate their own knowledge about the contents of the Smarties box from the knowledge of another person. They grasped the distinction between the external world as *they* perceived it, and the world as perceived by others. The psychologists then conducted the same experiment with three-year-olds, and the exact opposite result came back. The children consistently assumed that the grown-up would expect to find pencils in the box, not candy. They had not yet developed the faculty for building models of other people's mental states—they were trapped in a kind of infantile omniscience, where the knowledge you possess is shared by the entire world. The idea of two radically distinct mental states, each containing different information about the world, exceeded the faculties of the three-year-old mind, but it came naturally to the four-year-olds.

Our closest evolutionary cousins, the chimpanzees, share our aptitude for mind reading. The Dutch primatologist Frans de Waal tells a story of calculating sexual intrigue in his engaging, novel-like study, *Chimpanzee Politics.* A young, low-ranking male (named, appropriately enough, Dandy) decides to make a play for one of the females in the group. Being a chimpanzee, he opts for the usual chimpanzee method of expressing sexual attraction, which is to sit with your legs apart within eyeshot of your *objet de désir* and reveal your erection. (Try that approach in human society, of course, and you'll usually end up with a restraining order.) During this particular frisky display, Luit, one of the high-ranking males, happens upon the "courtship" scene. Dandy deftly uses his hands to conceal his erection so that Luit can't see it, but the female chimp can. It's the chimp equivalent of the adulterer saying, "This is just our little secret, right?"

De Waal's story—one of many comparable instances of primate intrigue—showcases our close cousins' ability to model the mental states of other chimps. As in the Smarties study, Dandy is performing a complicated social calculus in his concealment strategy: he wants the female chimp to know that he's enamored of her, but wants to hide that information from Luit. That kind of thinking seems natural to us (because it is!), but to think like that you have to be capable of modeling the contents of other primate minds. If Dandy could speak, his summary of the situation might read something like this: she knows what I'm thinking; he doesn't know what I'm thinking; she knows that I don't want him to know what I'm thinking. In that crude act of concealment, Dandy demonstrates that he possesses a gift for social imagination missing in 99.99 percent of the world's living creatures. To make that gesture, he must somewhere be aware that the world is full of imperfectly shared information, and that other individuals may have a perspective on the world that differs from his. Most important (and most conniving), he's capable of exploiting that difference for his own benefit. That exploitation—a furtive pass concealed from the alpha male—is only possible because he is capable of building theories of other minds.

Is it conceivable that this skill simply derives from a general increase in intelligence? Could it be that humans and their close cousins are just smarter than all those other species who flunk the mind-reading test? In other words, is there something specific to our social intelligence, something akin to a module hardwired into the brain's CPU—or is the theory of minds just an idea that inevitably occurs to animals who reach a certain threshold of general intelligence? We are only now beginning to build useful maps of the brain's functional topography, but already we see signs that "mind reading" is more than just a by-product of general intelligence. Several years ago, the Italian neuroscientist Giaccamo Rizzollati discovered a region of the brain that may well prove to be integral to the theory of other minds. Rizzollati was studying a section of the ventral premotor area of the monkey brain, a region of the frontal lobe usually associated with muscular control. Certain neurons in this field fired when the monkey performed specific activities, like reaching for an object or putting food in its mouth. Different neurons would fire in response to different activities. At first, this level of coordination suggested that these neurons were commanding the appropriate muscles to perform certain tasks. But then Rizzollati noticed a bizarre phenomenon. The same neurons would fire when the monkey observed another monkey performing the task. The pound-your-fist-on-the-floor neurons would fire every time the monkey saw his cellmate pounding his fist on the floor.

Rizzollati called these unusual cells "mirror neurons," and since his announcement of the discovery, the neuroscience community has been abuzz with speculation about the significance of the "monkey see, monkey do" phenomenon. It's conceivable that mirror neurons exist for more subtle, introspective mental states—such as desire or rage or tedium—and that those neurons fire when we detect signs of those states in others. That synchronization may well be the neurological root of mind reading, which would mean that our skills were more than just an offshoot of general intelligence, but relied instead on our brains' being wired a specific way. We know already that specific regions are devoted to visual processing, speech, and other cognitive skills. Rizzollati's discovery suggests that we may also have a module for mind reading.

The modular theory is also supported by evidence of what happens when that wiring is damaged. Many neuroscientists now believe that autistics suffer from a specific neurological disorder that inhibits their ability to build theories of other minds—a notion that will instantly ring true for anyone who has experienced the strange emotional distance, the radical introversion, that one finds in interacting with an autistic person. Autism, the argument goes, stems from an inability to project outside one's own head and imagine the mental life of others. And yet autistics regularly fare well on many tests of general intelligence and often display exceptional talents at math and pattern recognition. Their disorder is not a disorder of lowered intellect. Rather, autistics lack a particular skill, the way others lack the faculty of sight or hearing. They are mind blind.

Still, it can be hard to appreciate how rare a gift our mind reading truly is. For most of us, that we are aware of other minds seems at first blush like a relatively simple achievement—certainly not something you'd need a special cognitive tool for. I know what it's like inside my head, after all—it's only logical that I should imagine what's inside someone else's. If we're already self-aware, how big a leap is it to start keeping track of other selves?

This is a legitimate question, and like almost any important question that has to do with human consciousness, the jury is still out on it. (To put it bluntly, the jury hasn't even been convened yet.) But some recent research suggests that the question has it exactly backward—at least as far as the evolution of the brain goes. We're conscious of our own thoughts, the argument suggests, only because we first evolved the capacity to imagine the thoughts of others. A mind that can't imagine external mental states is like that of a three-year-old who projects his or her own knowledge onto everyone in the room: it's all pencils, no Smarties. But as philosophers have long noted, to be self-aware means recognizing the limits of selfhood. You can't step back and reflect on your own thoughts without recognizing that your thoughts are finite, and that other combinations of thoughts are possible. We know both that the pencils are in the box, *and* that newcomers will still expect Smarties. Without those limits, we'd certainly be aware of the world in some basic sense—it's just that we wouldn't be aware of *ourselves,* because there'd be nothing to compare ourselves to. The self and the world would be indistinguishable.

The notion of being aware of the world and yet not somehow self-aware seems like a logical impossibility. It feels as if our own selfhood would scream out at us after a while, "Hey, look at me! Forget about those Smarties—I'm thinking here! Pay attention to me!" But without any recognition of other thoughts to measure our own thoughts against, our own mental state wouldn't even register as something to think about. It may well be that self-awareness only jumps out to us because we're naturally inclined to project into the minds of others. But in a mind incapable of imagining the contents of other minds, that self-reflection wouldn't be missed. It would be like being raised on a planet without satellites, and missing the moon.

We all have a region of the retina where the optic nerve connects the visual cortex to the back of the retina. No rods or cones are within this area, so the corresponding area of our visual field is incapable of registering light. While this blind spot has a surprisingly large diameter (about six degrees across), its effects are minimal because of our stereo vision: the blind spots in each eye don't overlap, and so information from one eye fills in the information lacking in the other. But you can detect the existence of the blind spot by closing one eye and focusing the other on a specific word in this sentence. Place your index finger over the word, and then slowly move your finger to the right, while keeping your gaze locked on the word. After a few inches, you'll notice that the tip of your finger fades from view. It's an uncanny feeling, but what's even more uncanny is that your visual field suffers from this strange disappearing act anytime you close one eye. And yet you don't notice the absence at all—there's no sense of information being lost, no dark patch, no blurriness. You have to do an elaborate trick with your finger to notice that something's missing. It's not the lack of visual information that should startle us; it's that we have such a hard time noticing the lack.

The blind spot doesn't jump out at us because the brain isn't expecting information from that zone, and there's no other signal struggling to fill in the blanks for us, or pointing out that there is a blank in the first place. As the philosopher Daniel Dennett describes it, there are no centers of the visual cortex "responsible for receiving reports from this area, so when no reports arrive, there is no one to complain. An absence of information is not the same as information about an absence." We're blind to our blindness.

Perhaps the same goes with the theory of other minds. Without that awareness of other mental states reminding us of our own limitations, we might well be aware of the world, yet unaware of our own mental life. The lack of self-awareness wouldn't jump out at us for the same reason that the blind spot remains invisible: there's no feedback mechanism to sound the alarm that something's missing. Only when we begin to speculate on the mental life of others do we discover that we have a mental life ourselves.

If self-awareness is a by-product of our mind-reading skills, what propelled us to start building those theories of other minds in the first place? That answer comes more easily. The battle of nature-versus-nurture may have many skirmishes to come, but by now only the most blinkered anti-essentialist disagrees with the premise that we are social animals by nature. The great preponderance of human populations worldwide—both modern and "primitive"—live in extended bands and form complex social systems. Among the apes, we are an anomaly in this respect: only the chimps share our compulsive mixed-sex socializing. (Orangutans live mostly solitary lives; gibbons as isolated couples; gorillas travel in harems dominated by a single male.) That social complexity demands formidable mental skills: instead of outfoxing a single predator, or caring for a single infant, humans mentally track the behavior of dozens of individuals, altering their own behavior based on that information. Some evolutionary psychologists believe that the extraordinary expansion of brain size between *Homo habilis* and *Homo sapiens* (brain mass trebled over the 2-million-year period that separates the two species) was at least in part triggered by an arms race between Pleistocene-era extroverts. If successfully passing on your genes to another generation depended on a nuanced social intelligence that competed with other social intellects for reproductive privileges, then it's not hard to imagine natural selection generating a Machiavellian mental toolbox in a surprisingly short period.

The group element may even explain the explosion in sheer cranial size: social complexity is a problem that *scales* well—build a module that can analyze one person's mind, and all you need to do is throw more resources at the problem, and you can analyze a dozen minds with the same tools. The brain didn't need to invent any complicated new routines once it figured out how to read a single mind—it just needed to devote more processing power. That power came in the form of brain mass: more neurons to model the behavior of other brains, which themselves contained more neurons, for the same reason. It's a classic case of positive feedback, only it seems to have run into a ceiling of 150 people, according to the latest anthropological studies. We have a natural gift for building theories of other minds, so long as there aren't too many of them.

Perhaps if human evolution had continued on for another million years or so, we'd all be mentally modeling the behavior of entire cities. But for whatever reason, we stopped short at 150, and that's where we remained—until the new technologies of urban living pushed our collectivities beyond the magic number. Those oversize communities appeared too quickly for our minds to adapt to them using the tools of natural selection, and so we hit upon another solution, one engineered by the community itself, and not by its genes. We started building neighborhoods, groups within groups. When our lived communities extended beyond the ceiling of human comprehension, we started building new floors.

Mirror neurons and mind reading have an immense amount to teach us about our talents and limitations as a species, and there's no doubt we'll be untangling the "theory of other minds" for years to come. Whatever the underlying mechanism turns out to be, the faculty of mind reading—and its close relation, self-awareness—is clearly an emergent property of the brain's neural networks. We don't know precisely how that higher-level behavior comes into being, but we do know that it is conjured up by the local, feedback-heavy interactions of unwitting agents, by the complex adaptive system that we call the human mind. No individual neuron is sentient, and yet somehow the union of billions of neurons creates self-awareness. It may turn out that the brain gets to that self-awareness by first predicting the behavior of neurons residing in other brains—the way, for instance, our brains are hardwired to predict the behavior of light particles and sound waves. But whichever one came first—the extroverted chicken or the self-aware egg—those faculties are prime examples of emergence at work. You wouldn't be able to read these words, or speculate about the inner workings of your mind, were it not for the protean force of emergence.

But there are limits to that force, and to its handiwork. Natural selection endowed us with cognitive tools uniquely equipped to handle the social complexity of Stone Age groups on the savannas of Africa, but once the agricultural revolution introduced the first cities along the banks of the Tigris-Euphrates valley, the *Homo sapiens* mind naturally recoiled from the sheer scale of those populations. A mind designed to handle the maneuverings of less than two hundred individuals suddenly found itself immersed in a community of ten or twenty thousand. To solve that problem, we once again leaned on the powers of emergence, although the solution resided one level up from the individual human brain: instead of looking to swarms of neurons to deal with social complexity, we looked to swarms of individual humans. Instead of reverberating neuronal circuits, neighborhoods emerged out of traffic patterns. By following the footprints, and learning from their behavior, we built another ceiling on top of the one imposed on us by our frontal lobes. Managing complexity became a problem to be solved on the level of the city itself.

Over the last decade we have run up against another ceiling. We are now connected to hundreds of millions of people via the vast labyrinth of the World Wide Web. A community of that scale requires a new solution, one beyond our brains or our sidewalks, but once again we look to self-organization for the tools, this time built out of the instruction sets of software: Alexa, Slashdot, Epinions, Everything2, Freenet. Our brains first helped us navigate larger groups of fellow humans by allowing us to peer into the minds of other individuals and to recognize patterns in their behavior. The city allowed us to see patterns of group behavior by recording and displaying those patterns in the form of neighborhoods. Now the latest software scours the Web for patterns of online activity, using feedback and pattern-matching tools to find neighbors in an impossibly oversize population. At first glance, these three solutions—brains, cities, and software—would seem to belong to completely different orders of experience. But as we have seen over the preceding pages, they are all instances of self-organization at work, local interactions leading to global order. They exist on a continuum of sorts. The materials change as you jump from the scale of a hundred humans to a million to 100 million. But the system remains the same.

Amazingly, this process has come full circle. Hundreds of thousands—if not millions—of years ago, our brains developed a feedback mechanism that enabled them to construct theories of other minds. Today, we are beginning to create software applications that are capable of developing a theory of *our* minds. All those fluid, self-organizing programs tracking our tastes and interests, and measuring them against the behavior of larger populations—these programs are the beginning of a progression that will, in a matter of years, lead to a world where we regularly interact with media that seems to know us in some fundamental way. Software will recognize our habits, anticipate our needs, adapt to our changing moods. The first generation of emergent software—programs like SimCity and StarLogo—displayed a captivatingly organic quality; they seemed more like life-forms than the sterile instruction sets and command lines of early code. The next generation will take that organic feel one step further: the new software will use the tools of self-organization to build models of our own mental states. These programs won't be self-aware, and they won't pass any Turing tests, but they will make the media experiences we've grown accustomed to seem autistic in comparison. They will be mind readers.

From a certain angle, this is an old story. The great software revolution of the seventies and eighties—the invention of the graphic interface—was itself predicated on a theory of other minds. The design principles behind the graphic interface were based on predictions about the general faculties of the human perceptual and cognitive systems. Our spatial memory, for instance, is more powerful than our textual memory, so graphic interfaces emphasize icons over commands. We have a natural gift for associative thinking, thanks to the formidable pattern-matching skills of the brain's distributed network, so the graphic interface borrowed visual metaphors from the real world: desktops, folders, trash cans. Just as certain drugs are designed specifically as keys to unlock the neurochemistry of our gray matter, the graphic interface was designed to exploit the innate talents of the human mind and to rely as little as possible on our shortcomings. If the ants had been the first species to invent personal computers, they would have no doubt built pheromone interfaces, but because we inherited the exceptional visual skills of the primate family, we have adopted spatial metaphors on our computer screens.

To be sure, the graphic interface's mind-reading talents are ruthlessly generic. Scrolling windows and desktop metaphors are based on predictions about *a* human mind, not *your* mind. They're one-size-fits-all theories, and they lack any real feedback mechanism to grow more familiar with your particular aptitudes. What's more, their predictions are decidedly the product of top-down engineering. The software didn't learn on its own that we're a visual species; researchers at Xerox-PARC and MIT already knew about our visual memory, and they used that knowledge to create the first generation of spatial metaphors. But these limitations will soon go the way of vacuum tubes and punch cards. Our software will develop nuanced and evolving models of our individual mental states, and that learning will emerge out of a bottom-up system. And while this software will deliver information tailored to our interests and appetites, its mind-reading skills will be far less insular than today's critics would have us believe. You may read something like the *Daily Me* in the near future, but that digital newspaper will be compiled by tracking the interests and reading habits of millions of other humans. Interacting with emergent software is already more like growing a garden than driving a car or reading a book. In the near future, though, you'll be working alongside a million other gardeners. We will have more powerful personalization tools than we ever thought possible—but those tools will be created by massive groups scattered all across the world. When Patti Maes first began developing recommendation software at MIT in the early nineties, she called it collaborative filtering. The term has only grown more resonant. In the next few years, we will have personalized filters beyond our wildest dreams. But we will also be *collaborating* on a scale rivaled only by the cities we first started building six thousand years ago.

Those collaborations will build more than just music-recommendation tools and personalized newspapers. Our new ability to capture the power of emergence in code will be closer to the revolution unleashed when we figured out how to distribute electricity a century ago. Almost every region of our cultural life was transformed by the power grid; the power of self-organization—coupled with the connective technology of the Internet—will usher in a revolution every bit as significant. Applied emergence will go far beyond simply building more user-friendly applications. It will transform our very definition of a media experience and challenge many of our habitual assumptions about the separation between public and private life. A few decades from now, the forces unleashed by the bottom-up revolution may well dictate that we redefine intelligence itself, as computers begin to convincingly simulate the human capacity for open-ended learning. But in the next five years alone, we'll have plenty of changes to keep us busy. Our computers and television sets and refrigerators won't be thinking themselves, but they'll have a pretty good idea what we're thinking about.

Technology analysts never tire of reminding us that pornography is the ultimate early adopter. New technologies, in other words, are assimilated by the sex industries more quickly than by the mainstream—it was true for the printing press, for the VCR, for Web-based broadband. But video games are challenging that old adage. Because part of their appeal lies in their promise of new experiences, and because their audience is willing to scale formidable learning curves in pursuit of those new experiences, games often showcase cutting-edge technology before the tech makes its way over to the red-light district. Certainly that has been the case with emergent software. Gamers have been experimenting with self-organizing systems at least since SimCity's release in 1990, but the digital porn world remains, as it were, a top-down affair—despite the hype about putatively "interactive" DVDs.

In fact, video-game culture is the one arena today where you can see the "theory of other minds" integrated into a genuinely engaging media experience. Play any advanced first-person-shooter such as Quake or Unreal against computer opponents and you'll witness astonishingly lifelike behavior from the simulated gunslingers battling against you. They'll learn to anticipate your idiosyncrasies as a player; they'll form complicated flocking patterns with other computer "bots"; they'll grow familiar with new environments as they explore them. There's not much art to these talents, since they are mostly in service of blowing things up, but there is an undeniable intelligence to those computer opponents—an intelligence that is only indirectly controlled by the games' original programmers.

Will Wright's games have historically been the first to embrace bottom-up routines, but even an advanced game like The Sims falls short of his own ambitions in this arena. The residents of Simsville may display remarkably lifelike personality and behavioral traits, but they are unlikely to spontaneously develop a new skill that Wright didn't program into the game originally. You'll see them fall in love or zone out in front of the television, but you won't see one start yodeling or become a serial killer unless a human programmer has specifically added that behavior to the system. But Wright's dream is to have Sims that do develop unique behavior on their own, Sims that exceed the imagination of their creators. "I've been fascinated with adaptive computing for some time now," he says. "There are some rather hard problems to overcome, however. Some of the most promising technologies seem to also be the most parallel, like genetic algorithms, neural networks. These systems tend to learn by accumulating experience over a wide number of individual cases." Think here of Danny Hillis's number-sorting program. Hillis did manage to coax an ingenious and unplanned solution from the software, but it took thousands of iterations (not to mention a Connection Machine supercomputer). No gameplayer wants to sit around waiting for his on-screen characters to finish their simulated evolution before they start acting naturally. "In a game like The Sims," Wright says, "learning in 'user time' might best be accomplished by giving the characters a form of hypothetical modeling. In other words they might constantly be running 'micro-simulations' in their little heads—simulating a subset of the main simulation—to find ways of improving their decision-making."

But that learning need not be limited to the fictional universe of the game itself. "Another possibility would be to give the game some sense of how much the user is engaged and having fun," Wright speculates. "If we could measure this in some way—perhaps by analyzing the input stream and comparing it to a historical user profile—then we could design the game to learn what you like and enjoy. Each copy of the game would learn and evolve to fit each individual player. Maybe you're getting bored with the original gameplay; the game would detect this and try adding new elements to the game, getting more radical each time, until it hits on something you like. It would then take this and continue to evolve and refine it in directions that you find entertaining." Introduce real feedback into the equation—beyond the simple input of joysticks and trackballs—and suddenly the genre grows more flexible, more other-minded. The game becomes much more like a live performer, adapting to its audience, punching up certain routines while toning others down. Wright's vision is a significant step beyond the "choose your own path" vision of hypertext fiction championed in the early nineties. The "author" isn't presenting the "reader" with a selection of prefab threads to follow; the reader's interests and inclinations generate entirely novel threads, to the extent that the rules of the game vary from player to player. The first-generation interactive narratives were finally all about *choosing* one of several sanctioned links, picking one path over the others. The future that Wright envisions will be about creating a new path—or eliminating paths altogether.

Could such a model be applied to television? Not in the sense of growing your own sitcom, or choosing the ending of *ER*—but rather in the sense of growing your own network of programming. In the summer of 2000, a national ad campaign began running on the major networks, starring, for probably the first time in TV history, an office full of television programmers. "Look at these guys," the voice-over says contemptuously as the camera swoops through a workspace bustling with suits, casually canceling sitcoms and flirting with their personal assistants. "Network TV programmers. They decide what we watch and when we watch it." The camera tracks through an office door and homes in on an executive leaning back at his desk, contemplating the view from his corner office. Suddenly, two burly guys in black T-shirts appear at the corners of the screen. They pull the head programmer out of his chair and unceremoniously toss him out the window while the voice-over notes, "Who needs 'em?"

This strangely hostile spot is part of an extended campaign for a "personal digital TV recorder" called TiVo. While the ad itself belongs to the gonzo marketing tradition usually associated with day-trading outfits, its message may be more profound—and prophetic—than its homicidal-slacker demeanor suggests. At first glance, TiVo (and its main competitor, Replay) looks like the ultimate in *Daily Me*–style personalization: the device is primarily a large hard disk that records television programming based on your requests. In this respect, you can think of TiVo as a VCR that has a really good user interface, and that doesn't bother with the clutter and inconvenience of tapes. Because TiVo and Replay can analyze the program listings that you'll find on your cable provider's channel guide, the device can create automated filters that make your old VCR's scheduling features look as sophisticated as a Mr. Coffee unit. You can tell TiVo to record every episode of *NewsRadio* that appears on any channel anytime, or to record any Steve McQueen movies playing this week. Because it's permanently recording the last thirty minutes of television you've watched, you can actually pause live events or rewind them. The Einsteins at TiVo still haven't figured out how to jump ahead two minutes into the future, but if you're watching a program that's previously been recorded, you can zap through the ads with a click of the remote.

The upshot of all this gadgetry is that when you sit down at your television set, the question becomes less "What's on right now?" and more "What's on my hard drive right now?" This is where TiVo proposes to chuck the network programmers out the proverbial window. If the suits at Rockefeller Center decide that *Frasier* should move to Tuesday nights, and *Will and Grace* should move to Thursday, what do you care anymore? All your TiVo needs to know is that you're a *Frasier* fan, and you can watch the show anytime you want. You create your own prime time; you decide what you watch, and when you watch it.

This is a genuinely useful innovation, and while it bears some similarity to the original features of the VCR, the improvements in instant access and navigation make it a different beast altogether. And yet, it's still a transfer of control that looks more like the original vision of interactivity: instead of the network programmers calling the shots, *you* call the shots. There's a transfer of power in that change, but there's nothing emergent about it. The TiVo device "knows" what you want to watch, and thus in some relatively limited way, it possesses a theory of your mind. But it only knows what you want to watch because you programmed it yourself.

But TiVo and Replay—and their descendants—will also fall under the sway of self-organization. In five years, not only will every television set come with a digital hard drive—all those devices will also be connected via the Web to elaborate, Slashdot-style filtered communities. Every program broadcast on any channel will be rated by hundreds of thousands of users, and the TiVo device will look for interesting overlap between your ratings and those of the larger community of television watchers worldwide. You'll be able to build a personalized network without even consulting the channel guide. And this network won't necessarily follow the ultrapersonalization model of the *Daily Me.* Using self-organizing filters like the ones already on display at Amazon or Epinions, clusters of like-minded TV watchers will appear online. You might find yourself joining several different clusters, sorted by different categories: retirement-home senior citizens; West Village residents; GenXers; lacrosse fanatics. Visit the channel guide for each cluster, and you'll find a full lineup of programming, stitched together out of all the offerings available across the spectrum.

Despite the prevailing conventional wisdom, the death of the network programmer does not augur the death of communal media experiences. If anything, our media communities will grow stronger because they will have been built from below. Instead of a closed-door decision on West Fifty-seventh Street rebranding CBS as the "Tiffany Network," a cluster of senior citizens will form organically, and its constituents will participate far more directly in deciding what gets top billing on the network home page. To be sure, our media communities will grow smaller than they were in the days of *All in the Family* and *Mary Tyler Moore*—but they'll be *real* communities, and not artificial ones conjured up by the network programmers. There will still be a demand for entertaining television content—perhaps even more of a demand than there is today. But it will be distributed over a wider pool of shows, and the networks won't be able to force that demand on us by positioning shows in prime-time spots. The shows themselves will remain top-down affairs—the clusters won't be choosing the ending of this week's *Frasier* by popular vote—but the networks those shows find themselves aligned with will come from below. They'll be created by footprints, not fiat.

In a world where mass entertainment is delivered to us on our timetable, and according to our personal desires, how can advertising possibly hope to find a place at the table? If my TiVo device is already smart enough to detect my Audrey Hepburn obsession and my penchant for old episodes of the *Ben Stiller Show,* surely it's smart enough to know that I have zero interest in Colgate ads. Already, there are software applications that strip out banner ads from Web sites. What's stopping our digital filters from eliminating advertising entirely from our screens?

The answer, in a word, is nothing. If you thought the recording industry was frightened by Napster, imagine the terror on Madison Avenue. At least with Napster, consumers are still listening to music, even if they're not paying for it. With devices like TiVo and the banner blockers appearing on the market, there's a real chance that the public will stop tolerating ads altogether.

What are the salesman of Madison Avenue to do? There are three primary routes they can pursue. The first is to continue down the illustrious path that began with *E.T.*'s Reese's Pieces: product placement. If consumers start zapping past your thirty-second spots to watch the real content, then build your ads into the content. This approach has the risk of triggering a kind of marketing arms race between advertisers and consumers: future sitcom stars may well look like Formula One race cars, decked out in a dozen corporate logos, while the software wizards dream up new dynamic filters to block those logos from view. Alternately, the advertising industry can borrow a page from the MTV playbook and strive to make the ads as engaging and as indistinguishable from traditional content as possible. Imagine a future where every day is like Super Sunday: TV watchers program their TiVos to skip the tedious game itself and capture only the ads.

There is a third way, however—one that doesn't chip away at the wall between advertising and content, and that provides a genuine service to the consumer. And that is to make advertising smarter by tapping into the same feedback routines and self-organizing clusters that the content providers use. If advertising too starts to be rated and dynamically linked like any other unit in the mediasphere's emergent system, then specific ads will naturally find their way to consumers likely to respond well to their message. Already, I receive automated messages from Amazon alerting me to new releases that match my user profile. At first glance, these e-mails look utterly indistinguishable from the worst kind of spam, cluttering my in-box with yet another "special deal." But because I have a long and informative purchase history with Amazon, and because patterns in that history are generating the alert, I find the messages that Amazon sends me completely useful, and I often find myself buying items that they recommend. Having gotten a taste of this personalized advertising, I actually find it frustrating that other vendors that I've purchased goods from don't use the same system.

Progressive critics invariably find something sinister in the notion of smart advertising: all those corporations tracking our interests, and serving up ads custom-tailored to our needs, based on some devious mind-reading algorithm. Once you look beyond the admittedly significant privacy issues involved, the resistance to bottom-up, smart advertising strikes me as being absurdly reactionary and shortsighted. Imagine, for the sake of argument, that the development of advertising technology had proceeded in the opposite direction, and we had lived through most of the twentieth century with personalized ads like the Amazon e-mail alerts, and only now, at the beginning of a new era, did the technology arise that enabled mass advertising. All the social critics now up in arms over smart ads would be even more appalled by the blank, impersonal fire hose of mass advertising being directed at them. "These ads don't have any idea what we're interested in as individuals! All they know about us is that we're part of some age demographic, or that we're in a specific income bracket. In the old days, the ads at least used to know something about me personally. But these new mass ads are an affront to our individuality." Of course, I wish there were fewer billboards in the world, and less spam in my in-box. But if we're going to live in a world with advertising—and particularly if that world expects its entertainment to be partially subsidized by advertising—then I'd much prefer to see smart ads than stupid ones.

What's preventing charlatan mind readers from spamming my in-box with fake personalized ads, or tweaking the hidden algorithm so that my interests happen to align with merchandise they need to move? It's a legitimate problem, but it can be solved, presuming that two things are in place. First, we need strong anti-spam regulation that ensures that you can get off any mailing list at any time with a single e-mail request. Second—and most important—smart advertising systems should *themselves* be rated by the user community, at consumer sites like Epinions. You get book recommendations based on your ratings of other books; you sign up for a book recommendation service based on the *service's* ratings. A huckster who rejiggered his filtering software so that the most expensive products were promoted would see his ratings decline, as consumers discovered that the smart ads weren't all that smart after all. Just as you can adjust the quality threshold for posts on Slashdot, you could do the same with personalized advertising: sign me up for the most highly rated services and ignore the rest. Who knows? In a few decades, we might not have a need for the Consumer Protection Agency anymore—not because the corporations finally won out over the state, but because the consumers learned to regulate themselves.

Most attempts to speculate on the state of the Web five years out focus on the endlessly scrutinized dream of "convergence": the holy trinity of video, audio, and text at long last ushered into your living room via the same delivery mechanism, at speeds sufficient to convey the latest Lucas epic or Eminem release at the quality we've come to expect from CDs and DVDs. Once we hit that threshold, the critics like to tell us, the traditional media universe will no longer obey its previous laws of gravity, and a new order will emerge. What that new order will look like is a matter of great dispute. Some see a "one nation under AOL Time Warner" scenario; others envision a Hobbesian melee where everyone who sells zeros and ones is suddenly in the entertainment business. While the analysts disagree on the specifics of life after the revolution, there's a general consensus that the rise of convergence will finally trigger that last storming of the network media barricades.

No one who has spent any time contemplating the tsunami discharged by Napster can say with a straight face that the arrival of genuine convergence won't transform the media landscape. Two years ago, music downloaded from the Web was basically unlistenable for anyone who'd experienced post-Victrola audio quality; as I write, Napster's worse-than-CD-quality MP3 downloads have the recording industry transfixed with abject fear. But convergence is not the entire story. From a consumer's perspective, convergence will mean that the ordered universe of media offerings—prime-time sitcoms, top-forty radio, summer blockbusters—will be shattered into a million options, conveyed by a thousand providers. Turning on your television will be like logging on to the Web today: an infinite collection of links will beckon you—if not through the front door, then at least a few doors down the chain. This decentralized process is already under way: today's breakaway Nielsen hits have half the audience reach of *The Cosby Show* or *M*A*S*H*—and only two-thirds of Americans are even wired for cable. Imagine what happens when there are a million channels; imagine what happens when *channel* isn't the right word anymore, and we're simply surfing a giant hard drive of every song, movie, or television show that's ever been recorded. That is the inevitable future that awaits us at the other end of convergence.

It's a future that looks a lot like the present. We've lived through a comparable period of information expansion since the mid-nineties. The information available online doubles every six months, and we all know the vertiginous climb that exponential growth produces. The information overload presented by the billion or so HTML pages has forced us to reach for new tools to manage that glut, tools that eliminate the need for centralized archivists or editors, tools that lean on the entire community of surfers for their problem-solving. The same crisis will confront the media providers of 2005: the crisis of overload. For fifty years, the television industry—and all its tributaries—has been predicated on the idea that running a show on a network at 8 P.M. on Thursday nights guarantees a massive audience. But in a future where everyone is running the equivalent of a million TiVos off their desktop home entertainment device, what use is 8 P.M. Thursday night? You watch what you want to watch *when* you want to watch it, remember? And what use is a network in an age of infinite connectivity—when every program ever made is only a few clicks away? Why bother pledging your allegiance to NBC's "must-see TV" when it's just as easy to find that *Buffalo Bill* rerun anytime your heart desires a little Dabney Coleman?

How will we find our way through that kind of a mediated anarchy? Maybe we'll simply grow accustomed to the noise and learn to live with a remote control that feels more like a slot machine than a traditional guide. Maybe one opportunistic company will swoop in and become our primary conduit to the media frontier, as AOL partially did with the Web. But I suspect the overall media system will end up reaching a different equilibrium point, somewhere between Roman ultracentralization and the scattered chaos of the Dark Ages. Out of the turbulence of media convergence, the hill

towns will appear. They'll be built out of patterns of loca
and they'll be in continuous flux. But they will give shap
would otherwise be an epic expanse of shapelessness. Th
tainment world will self-organize into clusters of shared interest, created by software that tracks usage patterns and collates consumer ratings. These clusters will be the television networks and the record labels of the twenty-first century. The HBOs and Interscopes will continue to make entertainment products and profit from them, but when consumers tune in to the 2005 equivalent of *The Sopranos,* they won't be tuning in to HBO to see what's on. They'll be tuning in to the "Mafia stories" cluster, or the "suburban drama" cluster, or even the "James Gandolfini fan club" cluster. All these groups—and countless others—will point back to *The Sopranos* episode, and HBO will profit from creating as large an audience as possible. But the prominence of HBO itself will diminish: the network that actually serves up the content will become increasingly like the production companies that create the shows—a behind-the-scenes entity, familiar enough to media insiders, but not a recognized consumer brand. You'll enjoy HBO's programming, but you'll feel like you belong to your clusters. And you'll be right to feel that way, because you'll have played an important role in making them a reality.

Think of the media world as a StarLogo simulation. It begins with a perfectly ordered grid, like an aerial view of Kansas farmland: each network has its lineup in place, each radio station has its playlist. And then the convergence wave washes across that world and eliminates all the borders. Suddenly, every miniseries, every dance remix, every thriller, every music video ever made, is available from anywhere, anytime. The grid shatters into a million free-floating agents, roaming aimlessly across the landscape like those original slime mold cells. All chaos, no order. And then, slowly, clusters begin to form, shapes emerging out of the shapelessness. Some clusters grow into larger entities—perhaps the size of small cable networks—and last for many years. Other clusters are more idiosyncratic, and fleeting. Some map onto the physical world ("inner-city residents"); some are built out of demographic categories ("senior citizens"); many appear based on patterns in our cultural tastes that we never knew existed, because we lacked the tools to perceive them ("Asian-American Carroll O'Connor fans"). These new shapes will be like the aggregations of slime mold cells that we first encountered at the very beginning of this book; they will be like the towns blossoming across Europe eight hundred years ago; they will be like the neighborhoods of Paris or New York City. They will be like those other shapes because they will be generated by the same underlying processes: pattern-matching, negative feedback, ordered randomness, distributed intelligence. The only difference is the materials they are made of: swarm cells, sidewalks, zeros and ones.

In the end, the most significant role for the Web in all of this will not involve its capacity to stream high-quality video images or booming surround sound; indeed, it's quite possible that the actual content of the convergence revolution will arrive via some other transmission platform. Instead, the Web will contribute the metadata that enables these clusters to self-organize. It will be the central warehouse and marketplace for all our patterns of mediated behavior, and instead of those patterns being restricted to the invisible gaze of Madison Avenue and TRW, consumers will be able to tap into that pool themselves to create communal maps of all the entertainment and data available online. You might actually have the bits for *The Big Sleep* sent to you via some other conduit, but you'll decide to watch it because the "Raymond Chandler fans" cluster recommended the film to you, based on your past ratings, and the ratings of millions of like-minded folks. The cluster will build a theory of your mind, and that theory will be a group project, assembled via the Web out of an unthinkable number of isolated decisions. Each theory and each cluster will be more specialized than anything we've ever experienced in the top-down world of mass media. These mind-reading skills will emerge because for the first time our patterns of behavior will be exposed—like the sidewalks we began with—to the shared public space of the Web itself.

That promises a genuine revolution in what it means to be a media consumer, but it also demands a comparable revolution in the way businesses work. No company has more thoroughly explored the commercial possibilities of clusters and bottom-up organization than the celebrated auction site eBay. Since its launch in 1995, the site has been a virtual laboratory for experiments with clusters and self-regulating feedback. The "news" on eBay is almost entirely generated by the users of the service, and by the collective behavior of specific groups of users. The top auctions, the highly rated buyers-and-sellers lists, the user feedback, the communities formed around specific categories like Stamp Collecting or Consumer Electronics, the regional filters, the lists of new offerings from people you've bought from before—all of these are attempts to make patterns of group behavior transparent to individual users, the way a city neighborhood makes comparable patterns visible to its residents. EBay's founder, Pierre Omidyar, originally created the site to enable his wife to trade Pez Dispensers with other Pez fanatics worldwide; six years later the site harbors thousands of similar microcommunities, united by shared interests. If eBay had restricted itself to showcasing the collector's items that happened to be in vogue that month—Beanie Babies or PlayStation 2—the results wouldn't have looked all that different from your traditional shopping mall. But they wisely allowed their site to splinter into thousands of smaller clusters, like little eddies in the group current of their customer activity. Skeptics used to argue that online auctions would never become a mainstream activity because the electronic medium would make it easy for scam artists to sell bogus merchandise. Those critics wildly underestimated the extent to which software can create self-regulating systems, systems that separate the scoundrels from the honest dealers, the way Slashdot's quality filters separated quality from crap. Every seller on eBay has a public history of past deals; scam one buyer with a fake or broken item, and your reputation can be ruined forever. Like the public safety of Jacob's sidewalks, the eBay population polices itself with almost unbelievable efficiency, which is why the site now attracts more than 30 million users. And unlike almost any other Web-based commerce site, eBay has been consistently profitable from its early days. The history of self-organizing clusters includes the silk and fine linens bought and sold on Florence's Por Santa Maria or London's Savile Row. But don't underestimate the significance of Pez.

Still, if eBay is a model for the way bottom-up systems can transform the relationship between buyer and seller, can the principles of emergence be usefully applied to the internal structure of organizations? Is it possible to build corporate systems that are more like ant colonies than command economies? Marketplaces—even those dominated by global megacorporations—tend to work in decentralized ways, but the internal structures of most corporations today rely on org charts that look more like feudal states than slime molds. The market may be bottom-up, but it is populated by chronically top-heavy agents. Decentralized production and development have done wonders for the world of Open Source software, where certain fundamental rights of ownership have been disavowed, but it remains a real question whether the more proprietary wing of late capitalism can model its internal organization after ant farms or neural nets. For one, the unpredictability of emergent systems makes them an ideal platform for book recommending or gameplaying, but no one wants a business that might spontaneously fire a phalanx of middle managers for no discernible reason. Controlled randomness is a brilliant recipe for city life and ant foraging, but it's harder to imagine selling shareholders on it as a replacement for the CEO. Software designers like Danny Hillis or Oliver Selfridge leaned on evolutionary techniques to rein in their systems and to force them toward specific goals. But evolution requires many parallel generations to do its handiwork; no investor wants to wait around while her investment breeds a long-term strategy out of a million random business plans.

Still, emergent systems can be brilliant innovators, and they tend to be more adaptable to sudden change than more rigid hierarchical models. Those qualities make the principles of bottom-up intelligence tantalizing ones for businesses struggling to keep up with the twenty-first-century rate of change. A number of companies, concentrated mostly in the high-tech industry, have experimented with neural-net-like organizational structures, breaking up the traditional system of insular and hierarchical departments and building a more cellular, distributed network of small units, usually about a dozen people in size. Units can assemble into larger clusters if they need to, and those clusters have the power to set their own objectives. The role of traditional senior management grows less important in these models—less concerned with establishing a direction for the company, and more involved with encouraging the clusters that generate the best ideas. Imagine a corporate system structured like the Slashdot quality filters: in a traditional company, the CEO composes the posts himself; in a Slashdot-style company, he's merely tweaking the algorithm that promotes or demotes posts based on their quality. The vision for the company's future comes from below, out of the ever-shifting alliances of smaller groups. Senior management simply provides the feedback mechanism—in the form of bonuses, options, or increased resources—ensuring that the most productive clusters thrive. CEOs still have a place in even the most distributed corporate structure, but they're no longer allowed to be pacemakers. The Australian software company TCG, the Taiwanese Acer Group, and Sun Microsystems have all implemented cellular techniques with positive results. There's even a management-theory journal devoted to these developing models. It is called, appropriately enough, *Emergence.*

If decentralized intelligence can transform the way businesses work, what can it do for politics? That many New Economy companies have been so quick to embrace the emergent worldview—in both their products and their internal structure—can sometimes make it seem as though emergence belongs squarely to the libertarian camp. Certainly the emphasis on local control and the resistance to command systems resonates with the Gingrichian call for anti-big-government devolution. But the politics of emergence are not so readily classified. The intelligence of ant colonies may be the animal kingdom's most compelling argument for the power of the collective, and you can think of "local knowedge" as another way of talking about grassroots struggle. The libertarian right likes to rail against the centralized authority of the state, but at least most politicians in the world today are democratically elected, unlike the executives of most multinational corporations. The public sector has no monopoly on top-down systems, and there's no reason why progressives shouldn't also embrace decentralized strategies, even if those same strategies are being explored by right-wing think tanks and dot-coms. In fact, the needs of most progressive movements are uniquely suited to adaptive, self-organizing systems: both have a keen ear for collective wisdom; both are naturally hostile to excessive concentrations of power; and both are friendly to change. For any movement that aims to be truly global in scope, making it almost impossible to rely on centralized power, adaptive self-organization may well be the only road available.

Nowhere are the progressive possibilities of emergence more readily apparent than in the anti-WTO protest movements, which have explicitly modeled themselves after the distributed, cellular structures of self-organizing systems. The Seattle protests of 1999 were characterized by an extraordinary form of distributed organization: smaller affinity groups representing specific causes—anti-Nike critics, anarchists, radical environmentalists, labor unions—would operate independently for much of the time, only coming together for occasional "spokescouncil" meetings, where each group would elect a single member to represent their interests. As Naomi Klein reported in *The Nation,* "At some rallies activists carry actual cloth webs to symbolize their movement. When it's time for a meeting, they lay the web on the ground, call out 'All spokes on the web,' and the structure becomes a street-level boardroom." To some older progressives, steeped in the more hierarchical tradition of past labor movements, those diverse "affinity groups" seemed hopelessly scattered and unfocused, with no common language or ideology uniting them. It's almost impossible to think of another political movement that generated as much public attention without producing a genuine leader—a Jesse Jackson or Cesar Chavez—if only for the benefit of the television cameras. The images that we associate with the antiglobalization protests are never those of an adoring crowd raising their fists in solidarity with an impassioned speaker on a podium. That is the iconography of an earlier model of protest. What we see again and again with the new wave are images of disparate groups: satirical puppets, black-clad anarchists, sit-ins and performance art—but no leaders. To old-school progressives, the Seattle protesters appeared to be headless, out of control, a swarm of small causes with no organizing principle—and to a certain extent they're right in their assessment. What they fail to recognize is that there can be power and intelligence in a swarm, and if you're trying to do battle against a distributed network like global capitalism, you're better off becoming a distributed network yourself.

That is not a reason to embrace pure anarchy, of course. Ant colonies do not have leaders in any real sense, but they do rely heavily on rules: how to read patterns in the pheromone trail, when to change from foraging to nest-building, how to respond to other ants, and so on. An ant colony without local rules has no chance of creating a higher-level order, no chance of creating a collective intelligence. The antiglobalization movements are only beginning to figure out the proper rules for engagement between different cells. The spokescouncils of Seattle were a promising start, but learning how to cluster takes time. Klein writes, "What emerged on the streets of Seattle and Washington was an activist model that mirrors the organic, interlinked pathways of the Internet." But as we've seen countless times over the preceding pages, even the Web itself—the largest and most advanced man-made self-organizing system on the planet—is only now becoming capable of true collective intelligence. By any measure, the Web's mind-reading skills are embryonic at best, because we are still tweaking the rules of the system, still fiddling with how adaptive and intelligent clusters can prosper online. And if the Web's collective intelligence is still in its infancy, think of how much room the new protest movements must have to grow. But thus far, their instincts have been sound ones. Beneath the window-smashing and the Rage Against the Machine concerts, the anti-WTO activists are doing something profound, even in these early days of their movement. They are thinking like a swarm.

7

See What Happens

For years mathematicians have puzzled over a classic brainteaser known as the "traveling salesman problem." Imagine you're a salesman who has to visit fifteen cities during a business trip—cities that are distributed semirandomly across the map. What is the shortest route that takes you to each city exactly once? It sounds like a simple enough question, but the answer is maddeningly difficult to establish. Even with the number of cities set at a relatively modest fifteen, billions of potential routes exist for our traveling salesman. For complicated reasons, the traveling salesman problem is almost impossible to solve definitively, and so historically mathematicians—and traveling salesmen, presumably—have settled for the next best thing: routes that are tolerably short, but not necessarily the shortest possible.

This might sound like an arcane issue, given the real-world decline of the traveling salesman, but the core elements of the problem lie at the epicenter of the communications revolution. Think of those traveling salesmen as bits of data, and the cities as Web servers and routers scattered all across the globe. Being able to calculate the shortest routes through that network would be a godsend for a massive distributed system like the Internet, where there may be thousands of "cities" on any given route, instead of just fifteen. The traveling salesman may finally have been killed off for good by online retailers like Amazon.com, but the traveling salesman *problem* has become even more critical to the digital world.

In late 1999, Marco Dorigo of the Free University of Brussels announced that he and his colleagues had hit upon a way of reaching "near-optimal" solutions to the traveling salesman problem that was notably more time-efficient that any traditional approach. Dorigo's secret: let the ants do the work.

Not literal ants, of course. As we saw at the beginning of this book, ant colonies have an uncanny ability to calculate the shortest path to different food sources, using their simple language of pheromone trails. Dorigo's insight was to solve the traveling salesmen problem the way an ant colony would: send out an army of virtual salesman to explore all possible routes on the map. When a salesman successfully completes a journey to all fifteen cities, he then traces his path back to the starting city, depositing a small amount of virtual "pheromone" along the way. Because the total amount of pheromone is finite, it is spread more thinly along longer routes, and more heavily along shorter ones. With thousands of ants exploring the map, some sections of shorter routes quickly accumulate thick layers of pheromone, while less efficient routes have almost no pheromone at all. After the first round of exploration, a new batch of virtual salesmen are released and encouraged to follow the routes that have been most heavily dosed with pheromones. After several repeated sessions, the salesmen swarm starts homing in on the shortest routes, reaching a near-optimal solution to the traveling salesman problem without using anything resembling traditional calculus or a central problem-solver. Since Dorigo's announcement of his results, France Telecom, British Telecommunications, and MCI have applied antlike routing strategies to their telephone and data networks. Early studies show that Dorigo's approach is significantly more efficient than the Open Shortest Path First routine used by the Internet to distribute data between nodes on the network. A few years from now, our online interactions may be sustained by the bottom-up power of swarm intelligence. And once again, the ants will have figured it out long before we did.

What kind of data will those future networks transmit? Soon after this book's publication, the Net will be teeming with the digital inhabitants of Will Wright's latest creation, The Sims Online. A fusion of The Sims and SimCity, the game allows players to collectively build cities as part of a massive network collaboration. Unlike either previous game, all the citizens of the world are controlled by actual human beings, logged into the system from all around the world. As in The Sims, you can zoom into your own character's living room, or visit a friend's house down the street for a neighborhood barbecue. But you can also zoom out to see the entire landscape that the players have created. An early draft of the game that Wright showed me included a brilliant neighborhood-creation system that seemed straight out of the pages of Jane Jacobs. City neighborhoods are defined from the bottom up, as players establish their own homesteads in various regions of the virtual space. Any player can create his or her own private neighborhood, the way they can create a name for their virtual character on the screen. But you can also persuade your neighbors to adopt your neighborhood name as well. When a certain number of citizens have declared their allegiance to a specific neighborhood, the system officially recognizes that district and gives it a special sign, along with various tax breaks. The bigger the neighborhood, the bigger the sign, and the more lucrative the benefits. It's a classic marriage of bottom-up growth and top-down management: let the neighborhoods come from below, but build incentives into the system to encourage their growth.

Our newfound access to virtual cities on the computer screen hasn't abated our appetite for real-world city living. Five years ago, most digital-savvy social critics predicted that the rise of the Web and various telecommuting appliances would deliver the death blow to city living, finishing a forty-year process that had begun with the surburban flight of the postwar years. We'd all be living on our Wyoming ranches in ten years, dialing into the office instead of straphanging on an overcrowded subway. Of all the lofty predictions of the midnineties, none have proven to be more misguided than those eulogies for urban life. The digital revolution has turned out to be a tremendous energizer for dense urban centers like San Francisco, New York, and Seattle—for reasons that date back to the guild system and trade clusters of twelfth-century Florence. Industries driven by ideas naturally gravitate toward physical centers of idea generation, even in an age of instant data transmissions. Bright minds with shared interests still flock together, even when they have wireless modems and broadband in their living rooms. Now smaller settlements are trying to learn from the street-centered dynamism of the traditional organic city: the New Urbanist movement has begun to transform America's suburban development practices, by following the rules outlined by Jane Jacobs almost a half century ago: shorter blocks, livelier sidewalks, mixed-used zoning, and pedestrian-based transportation. Despite the Bengali typhoon of the digital revolution—or perhaps, in part, because of it—the old-style self-organizing city is today as vital and relevant as it has ever been.

The vitality has its downsides, of course: rents, congestion, traffic jams. Even in the most sidewalk-centric cities, the flow of automobiles through the complex latticework of streets poses an organizational problem that rivals the traffic of information across the World Wide Web. For decades, urban engineers have built ever-more-complicated systems to direct the paths of automobiles through congested city streets, by observing the patterns of traffic and tweaking the stoplights and street directions where problems arose. But city traffic is a problem of organized complexity, and it is best tackled with bottom-up solutions, not top-down ones. Almost fifty years after he first came up with the idea for Pandemonium, Oliver Selfridge has embarked on a quest for exactly that solution, building a learning network of traffic lights that will find an optimal system in changing conditions. Selfridge wants to attack the problem of traffic the way Danny Hillis attacked the problem of number sorting: by giving the network the general goal of minimizing delays, but letting the overall system figure out the details, using the tools of feedback, neighbor interaction, and pattern recognition that are the hallmark of all self-organizing systems. Traffic jams themselves are a particularly crude form of emergent behavior, and for years we've been battling them with engineering solutions. Selfridge wants to take the master planners out of the equation. Make the traffic lights smart—by connecting them and feeding them information about backups or accidents—and you have a solution that can actually manage the immense and constantly changing problem of urban movement. You can conquer gridlock by making the grid itself smart.

What connects these developments? Imagine a kind of conceptual tracking shot of life two or three years from now, a movement from scale to scale—like the wonderful Charles and Ray Eames film, *Powers of Ten,* which starts with a view of the Milky Way and steadily zooms all the way to a person lying in a park in Chicago, and then all way to the subatomic particles contained within that person's hand.

Only in our long zoom do we find, at each scale, the same behavior repeating itself again and again. Begin on the scale of the city itself, its neighborhoods pulsing and thriving, as they have for centuries, sending signals out to the world, and drawing human beings into those neighborhoods, like massive global magnets. The flow of people through the city is now regulated by an intelligent traffic network, evolving and learning in response to patterns of automobile movement. You or I live in one of those immense systems, contributing to its continued development the way a single slime mold cell contributes to the larger aggregation—and as part of our life in that city we entertain ourselves by simulating the self-organizing energy of city life by playing a game on our computer screen, building virtual neighborhoods collectively with thousands of other networked players all across the world. On the scale of the city, and the scale of the screen, our lives embrace the powers of emergence.

Now zoom in another level, to the individual bits of information that convey our virtual city-building to our networked compatriots. These too find their way across the infosphere by drawing on the distributed logic of swarm behavior, building their complex itineraries from below. The network is smart, but its intelligence is the intelligence of an ant colony, not a centralized state. And how did these new smart networks come into being? Drop down one more level on the chain, to the neural networks of the human mind, and their extraordinary aptitude for pattern recognition. The mind of a researcher in Brussels sees a connection between the collective behavior of ant colonies and the routing problems endemic to large-scale information networks—sees the connection because his brain contains a marvelously agile device for detecting shared patterns in disparate fields. That device runs on its own kind of swarm logic, with no central office in command. One kind of decentralized intelligence (the human brain) grasps a new way to apply the lessons of another decentralized intelligence (the ants), which then serves as a platform (the network) for the transmission of another kind (the virtual cities), which we enjoy while sitting safely in our apartments in the neighborhoods of the planet's largest man-made self-organizing system (the real city). It is emergence all the way down the chain.

Can that chain be extended in new directions—both on the atomic scale of digital information and the macroscale of collective movements? Will computers—or networks of computers—become self-aware in the coming years, by drawing upon the adaptive open-endedness of emergent software? Will new political movements or systems explicitly model themselves after the distributed intelligence of the ant colony or the city neighborhood? Is there a fourth stage in the developing web of emergence that takes us beyond the mind readers into something even more lifelike? Is there a genuine global brain in our future, and will we recognize ourselves in it when it arrives?

Certainly the world has never been better prepared for these developments to become reality; if we don't enter the fourth phase of emergence in the coming decades, it won't be for lack of trying. But it is both the promise and the peril of swarm logic that the higher-level behavior is almost impossible to predict in advance. You never really know what lies on the other end of a phase transition until you press play and find out. That is the lesson of Gerald Edelman's recipe for simulating a flesh-and-blood organism: you set up a system of various pattern-recognition devices and feedback loops, connecting the virtual organism to a simulated environment. And then you see what happens.

Even the most optimistic champions of self-organization feel a little wary about the lack of control in such a process. But understanding emergence has always been about giving up control, letting the system govern itself as much as possible, letting it learn from the footprints. We have come far enough in that understanding to build small-scale systems for our entertainment and edification, and to appreciate more thoroughly the emergent behavior that already exists at every scale of our lived experience. Are there new scales to conquer, new revolutions that will make the top-down revolutions of the industrial age look minor by comparison? On the hundred-year scale, or the scale of millennia, there may be no question more interesting, and no question harder to answer.

NOTES

11 Without any apparent: "When Slime Is Not So Thick," BBC News, August 27, 2000.

13 Anyone who has ever: Indeed, we know now that we are closer to the slime mold colonies that we initially thought: "Bacteria of several different kinds got together, more than a thousand million years ago, to form the 'eucaryotic cell.' This is our kind of cell, with a nucleus and other complicated internal parts, many of them put together from intricately folded internal membranes, like the mitochondria which I briefly pointed to in Figure 5.2. The eucaryotic cell is now seen as derived from a colony of bacteria. Eucaryotic cells themselves later got together into colonies." Dawkins, 1996, 286–87.

13 If we could: "It is, in fact, scarcely more than a philosophical anticipation of the cell theory, according to which most of the animals and plants of moderate size and all of those of large dimensions are made up of units, cells, which have many if not all the attributes of independent living organisms. The multicellular organisms may themselves be the building bricks of organisms of a higher stage, such as the Portuguese man-of-war, which is a complex structure of differentiated coelenterate polyps, where the several individuals are modified in the different ways to serve the nutrition, the support, the locomotion, the excretion, the reproduction, and the support of the colony as a whole." Wiener, 155.

13 "I was at": Interview with Evelyn Fox Keller, conducted July 2000.

13 While the field: "Alan was familiar with Schrödinger's 1943 lecture, 'What Is Life,' which deduced the crucial idea that genetic information must be stored at molecular level, and that the quantum theory of molecular bonding could explain how such information could be preserved for thousands of millions of years. At Cambridge, Watson and Crick were busy in the race against their rivals to establish whether this was really so, and how. But the Turing problem was not that of following up Schrödinger's suggestion, but that of finding a parallel explanation of how, granted the production of molecules by the genes, a chemical soup could possibly give rise to a biological pattern. He was asking how the information in the genes could be translated into action. Like Schrödinger's contribution, what he did was based on mathematical and physical principle, not on experiment; it was a work of scientific imagination." Hodges, 431.

14 Turing's paper had: "Before the war [Turing] had read the classic work *Growth and Form* by the biologist D'Arcy Thompson, published in 1917 but still the only mathematical discussion of biological structure. He was particularly fascinated by the appearance in nature of the Fibonacci numbers—the series beginning

1, 1, 2, 3, 5, 8, 13, 21, 34, 55, 89 . . .

in which each term was the sum of the previous two. They occurred in the leaf arrangement and flower patterns of many common plants, a connection between mathematics and nature which to others was a mere oddity, but to him deeply exciting." Ibid., 207.

15 What if there: Evelyn Fox Keller, "The Force of the Pacemaker Concept in Theories of Aggregation in Cellular Slime Mold," in *Reflections on Gender and Science* (Yale University Press, 1996).

16 Indeed, the pacemaker: Resnick, 122.

16 reigning model: "[The Keller-Segel model] received very little attention, and no critique ever appeared. Instead, attention shifted away from field-theoretic models toward models of the individual cell. In part, this shift of interest may have been due to the failure of the Keller-Segel model to predict wavelike global oscillation of the aggregation field. . . . But even more, the shift was probably due to a philosophical antipathy to holistic models." Garfinkel, 187.

17 You can think: "The process of slime-mold aggregation is now viewed as one of the classic examples of self-organizing behavior." Ibid., 51.

17 It also unearthed: In *Death and Life of Great American Cities,* Jane Jacobs describes the decentralizing mentality this way: "In principle, these are much the same tactics as those that have to be used to understand and to

help cities. In the case of understanding cities, I think the most important habits of thought are these:

"1. To think about processes;
"2. To work inductively, reasoning from particulars to the general rather than the reverse;
"3. To seek for 'unaverage' clues involving very small quantities, which reveal the way larger and more 'average' quantities are operating." P. 440.

18 Marvin Minsky in: Marvin Minsky, *The Society of Mind.*
18 In a more technical: Ilya Prigogine and G. Nicolis, *Exploring Complexity.*
20 By watching the: "Understanding something in just one way is a rather fragile kind of understanding. Marvin Minsky has said that you need to understand something at least two different ways in order to really understand it. Each way of thinking about something strengthens and deepens each of the other ways of thinking about it. Understanding something in several different ways produces an overall understanding that is richer and of a different nature than any one way of understanding." Resnick, 103.
21 Self-organization became: The Santa Fe Institute is of course most famous for its work in the related field of "chaos theory." "As noted by Farmer and Packard (1986), the study of self-organizing systems is, in some ways, the 'related opposite' of the study of chaos: in self-organizing systems, orderly patterns emerge out of lower-level randomness; in chaotic systems, unpredictable behavior emerges out of lower-level deterministic rules." Ibid., 14.
21 But in the: Jane Jacobs describes these systems as "dynamically stable systems": "Every kind of system that is neither inert nor disintegrated. This includes all living systems: ecosystems, organisms, cells composing organisms, microorganisms. It also includes many inanimate systems: rivers, the atmosphere, the crust of the earth. Human settlements, business enterprises, economies, governments, nations, civilizations—they're all dynamically stable systems. Jacobs, 2000, 85.
22 The first section: "The dynamics of ant colony life has some features in common with many other complex systems: Fairly simple units generate complicated global behavior. If we knew how an ant colony works, we might understand more about how all such systems work, from brains to ecosystems. Because we don't yet comprehend any natural complex system, I think it is premature to say how general a theory we may eventually achieve. The intriguing question about task allocation is how might an ant react to local events, in a simple way, that in the aggregate produces colony

behavior? The same kinds of questions come up, over and over, throughout biology: How do neurons respond to each other in a way that produces thoughts? How do cells respond to each other in a way that produces the distinct tissues of a growing embryo? How do species interact to produce predictable changes, over time, in ecological communities? These are the big, general questions of biology, and many of us dream that when we have the answers, from different fields of biology, it will be possible to see similar processes at work from cells to ecosystems." Gordon, 141–42.

22 The epic clash: Norbert Wiener makes a parallel observation in *Cybernetics*: "This desire to produce and to study automata has always been expressed in terms of the living technique of the age. In the days of magic, we have the bizarre and sinister concept of the Golem, that figure of clay into which the Rabbi of Prague breathed life with the blasphemy of the Ineffable Name of God. In the time of Newton, the automaton becomes the clockwork music box, with the little effigies pirouetting stiffly on top. In the nineteenth century, the automaton is a glorified heat engine, burning some combustible fuel instead of the glycogen of the human muscles. Finally, the present automaton opens doors by means of photocells, or points guns to the place at which a radar beam picks up an airplane, or computes the solution to a different equation." Wiener, 39–40.

29 "And then we": Interview conducted with Gordon, September 1999.

31 "It would be": Gordon, 117.

31 The harvester ants: As the legendary biologist W. D. Hamilton argued in a famous paper from 1964, the social cohesion of ant colonies is interestingly tied to their genetics: while you share on average half of your genes with your siblings, the sister ants that populate a colony share three-quarters of their genes, due to a complicated process of sex determination in ant societies. Those shared genes imply a greater communal interest in preservation—even greater than the connection between parent and child. Being in the same "genetic boat" tends to lead toward greater cooperation: "Whether you are a bunch of genes or a bunch of memes, if you're all in the same boat you'll tend to perish unless you are conducive to productive coordination. For genes, the boat tends to be a cell or a multicelled organism or occasionally, as we'll see shortly, a looser grouping, such as a family; for memes, the boat is often a larger social group—a village, a chiefdom, a state, a religious denomination, Boy Scouts of America, whatever. Genetic evolution thus tends to create smoothly integrated organisms, and cultural evolution tends to create smoothly integrated groups of organisms." Wright, 257.

31 In other words: The distinction between the colony's being dependent on the queen, and being actually *controlled* by her, has been obscured by some

commentators drawing on insect societies as a way of thinking about human social organization. "Apart from human communities, war exists only among the social insects, which anticipated urban man in achieving a complex community of highly specialized parts.

"As far as external observations can show, one certainly does not find religion or ritual sacrifice in these insect communities. But the other institutions that accompanied the rise of the city are all present: the strict division of labor, the creation of a specialized military caste, the techniques of collective destruction, accompanied by mutilation and murder, the institution of slavery, and even, in certain species, the domestication of plants and animals. Most significant of all, the insect communities that exhibit these traits boast the institution I have taken to be central in this whole development: the institution of kingship. Kingship, or rather, its feminine equivalent, queenship, has been incorporated as a supreme biological fact in these insect societies; so that what is only a magic belief in early cities, that the life of the whole community depends on the life of the monarch, is an actual condition in insectopolis. On the queen's health, safety, and reproductive capacity the continued existence of the hive does in fact depend. Here and only here, does one find such organized collective aggression by a specialized military force as one finds first in the ancient cities." Mumford, 1961, 46.

35 This constitutes one: Information about the history of Manchester from Marcus, 5–6.

35 "From this foul": Quoted in Marcus, 15. "Considering this new urban area on its lowest physical terms, without reference to its social facilities or its culture, it is plain that never before in recorded history had such vast masses of people lived in such a savagely deteriorated environment, ugly in form, debased in content. The galley slaves of the Orient, the wretched prisoners in the Athenian silver mines, the depressed proletariat in the insulae of Rome—these classes had known, no doubt, a comparable foulness; but never before had human blight so universally been accepted as normal: normal and inevitable." Mumford, 1961, 474.

36 His three years: " . . . it is difficult to conceive of how Engels could have made this exceptionally profound and enduring investment of himself had there not been much in the past on which he could continue to rely. On the one hand, he was a young man intent upon burning his bridges behind him. On the other, he knew in some part of himself that those bridges were built of fireproof material. We can put it another way. It does not detract too much from the existential reality of his decision to say that he was jumping into the abyss with a parachute. Or perhaps it does; perhaps that is one of the ways of distinguishing between the qual-

ities of existential and historical choices in their classical modes, between the post-Hegelian Kierkegaard and the post-Hegelian Marx and Engels." Marcus, 128.

37 "Still . . . I cannot": "To their places of business in the center of the town by the shortest routes, which run right through all the working class districts, without even noticing how close they are to the most squalid misery which lies immediately about them on both sides of the road. This is because the main streets which run from the Exchange in all directions out of the city are occupied almost uninterruptedly on both sides by shops, which are kept by members of the middle and lower-middle classes." Engels, 86.

38 "It is indeed": "Their houses lie outside the working-class belt, and between that belt and the more favorably situated suburban estates of the upper middle class. And their shops and small businesses are located along the main thoroughfares—acting so to speak as insulators for the city's system of communication. Their location in space, therefore, is at both critical junctures an intermediate one. And their intermediary position is not merely structural but functional as well. They are acting as buffers between the antagonistic extremes." Marcus, 172–73.

39 "The method of": Quoted in Buck-Morss, 269.

41 There's no need: " . . .the Baron von Haussmann, in the course of building the boulevard Saint-Michel, that bleak, noisy thorough-fare, tore through the heart of the ancient Latin Quarter, which had been an almost autonomous entity since the Middle Ages. And he took the simplest of all methods of improving one portion of it: he wiped it out. He not merely cleared the area surrounding the Schools, but in a side-swipe even cut off part of the Gardens of the Palais by Luxembourg, sacrificing to straight lines, broad avenues and unimpeded vehicular traffic the specific historical character of the quarter and all the complex human needs and purposes it served. These baroque clichés of power, hardly even with the decency of a disguise, lingered right into the twentieth century: witness the plowing of the Seventh Avenue extension through the one historic quarter of New York that had integrity and character, or the similar, even more grandiose effacement created by the misconceived Benjamin Franklin Boulevard in Philadelphia—the latter a brutal gash from which the city has not recovered in more than thirty years." Mumford, 1961, 388.

41 The pattern is: Hodges, 428–29.

42 As part of: "The argument depended on the key being absolutely patternless, and spread evenly over the possible digits, for otherwise the analyst would have reason to prefer one guess or another. Indeed, discerning

a pattern in the apparently patternless was essentially the work of the cryptanalyst, as of the scientist." Ibid., 154.

43 That very morning: Ibid., 466.

44 Shannon and Turing: Ibid., 251.

45 But Shannon pushed: "Shannon had always been fascinated with the idea that a machine should be able to imitate the brain; he had studied neurology as well as mathematics and logic, and had seen his work on the differential analyzer as a first step towards a thinking machine. They found their outlook to be the same: there was nothing sacred about the brain, and that if a machine could do as well as a brain, then it would be thinking—although neither proposed any particular way in which this might be achieved. This was a back-room Casablanca, planning an assault not on Europe, but on inner space." Ibid.

46 Dense with equations: "We are beginning to see that such important elements as the neurons, the atoms of the nervous complex of our body, do their work under much the same conditions as vacuum tubes, with their relatively small power supplied from outside by the circulation, and that the bookkeeping which is most essential to describe their function is not one of energy. In short, the newer study of automata, whether in the metal or in the flesh, is a branch of communication engineering, and its cardinal notions are those of message, amount of disturbance or 'noise'—a term taken over from the telephone engineer—quantity of information, coding technique, and so on." Wiener, 42.

47 "This statistical method": Weaver, 66.

48 "They are all": Ibid., 69.

48 "The great central": Ibid., 71.

49 To solve the: " . . . unlike linear equations (the type most prevalent in science), nonlinear ones are very difficult to solve analytically, and demand the use of detailed numerical simulations carried out with the help of digital machines. This limitation of analytic tools for the study of nonlinear dynamics becomes even more constraining in the case of nonlinear combinatorics. In this case, certain combinations will display emergent properties, that is, properties of the combination as a whole which are far more than the sum of its individual parts. These emergent (or 'synergistic') properties belong to the interactions between parts, so it follows that a top-down analytical approach that begins with the whole and dissects it into its constituent parts (an ecosystem into species, a society into institutions), is bound to miss precisely those properties. In other words, analyzing a whole into parts and then attempting to model it by adding up the components will fail to capture any property that emerged from complex interactions, since the effect of the latter may be

multiplicative (e.g., mutual enhancement) not just additive." De Landa, 1997, 17–18.

50 Jacobs had just: "Then in the mid-1950's, Mr. Moses came up with a new plan for erosion. This once involved a major depressed highway cutting through the center of the park, as a link for carrying a heavy volume of high-speed traffic between midtown Manhattan and a vast, yawing Radiant City and expressway which Mr. Moses was cooking up south of the park." Jacobs, 1961, 360–61.

51 "This order is": Ibid., 50.

51 "We may wish": Ibid., 434.

51 Jacobs's book would: Although he disagreed with Jacobs on a number of fronts, Lewis Mumford was also using the language of emergence to describe city development around the same period: "The city came as a definite emergent in the paleo-neolithic community: as emergent in the definite sense that Lloyd Morgan and William Morton Wheeler used that concept. In emergent evolution, the introduction of a new factor does not just add to the existing mass, but produces an over-all change, a new configuration, which alters its properties. Potentialities that could not be recognized in the pre-emergent stage, like the possibility of organic life developing from relatively stable and unorganized 'dead' matter, then for the first time become visible." Mumford, 1961, 29.

52 "Vital cities have": Jacobs, 1961, 447–48.

52 And at MIT's: As usual, Turing was ahead of the game here: "It has been said that computing machines can only carry out the purposes that they are instructed to do. This is certainly true in the sense that if they do something other than what they were instructed then they have just made some mistake. It is also true that the intention in constructing these machines in the first instance is to treat them as slaves, giving them only jobs which have been thought out in detail, jobs such that the user of the machine fully understands in principle what is going on all the time. Up till the present machines have only been used in this way. But is it necessary that they should always be used in such a manner? Let us suppose we have set up a machine with certain initial instruction tables, so constructed that these tables might on occasion, if good reason arose, modify those tables. One can imagine that after the machine had been operating for some time, the instructions would have altered out of recognition, but nevertheless still be such that one would have to admit that the machine was still doing very worthwhile calculations." Quoted in Hodges, 358.

53 "Mostly my participation": Interview with Selfridge conducted October 2000.

54 "We are proposing": Selfridge, 1.
57 After a few: Levy, 155.
58 Holland's system revolved: Richard Dawkins has proposed that the analogy can run the other direction as well: "You can, if you wish, think of the genes in all the populations of the world as constituting a giant computer, calculating costs and benefits and currency conversions, with the shifting patterns of gene frequencies doing duty for the shuttling 1s and 0s of an electronic data processor. It is quite an illuminating insight . . ." Dawkins, 1996, 72.
60 "But I recognized": Interview with Jefferson conducted December 2000.
66 You can date: Kelly, 235.
67 Our minds may: Some researchers argue that the centralized mind-set is hardwired into our brain; in other words, we default to top-down explanations and only reconcile ourselves to bottom-up explanations after extensive training. "People also view the workings of the economy in centralized ways, assuming singular causes for complex phenomena. Children, in particular, seem to assume strong governmental control over the economy. (Of course, governments do play a large role in most economies, but children assume they play an even larger role than they actually do.) In interviews with Israeli children between eight and fifteen years old, psychologist David Leiser (1983) found that nearly half of the children assumed that the government sets all prices and pays all salaries. Even children who said that employers pay salaries often believed that the government provides the money for the salaries. A significant majority of the students assumed that the government pays the increased salaries after a strike. And many younger children had the seemingly contradictory belief that the government is also responsible for organizing strikes." Resnick, 123.
73 And while they: "Other ant and termite species are specialized to cultivate fungi underground, planting the spores, weeding the gardens to rid them of competing fungi species, and fertilizing them with compost mulched from chewed-up leaves. In the case of the famous leafcutter ants of the New World tropics, all the foraging efforts of their 8-million-strong colonies are directed towards harvesting fresh-cut leaves." Dawkins, 1996, 264–65.
73 massive environmental impact: "Ants farm fungi, raise aphids as livestock, launch armies into wars, use chemical sprays to alarm and confuse enemies, capture slaves. . . . They do everything but watch television." Thomas, 12.
73 They lack our: Wilson and Holldobler, 1.
74 Harvester ant colonies: " . . . a colony's soldiers give off an odor, a pheromone distinctive to the soldiers. If the odor falls below a certain

level in the colony, it means that the proportion of soldiers is less than normal; there's no mistaking the purport of the feedback. Since the drop automatically causes the nursery to supply more soldiers, there's no room for mistake in the response—or in cessation of the response, either. Pheromone feedback from newly produced soldiers reports, 'That's enough soldiers.' In short, data, the meaning of the data, and appropriate responses to the data are all perfectly integrated." Jacobs, 2000, 109.

75 "The sum of": Wilson and Holldobler, 227.

75 "Let's get rid": Ibid., 252.

79 Without those haphazard: "Randomness plays yet another role in some self-organizing processes—it makes possible the exploration of multiple options. Ant researcher Jean-Louis Deneubourg notes that ants do not follow pheromone trails perfectly. Instead, ants have a probabilistic chance of losing their way as they follow the trails. Deneubourg and his colleagues (1986) argue that this 'ant randomness' is not a defective stage on an evolutionary path 'towards an idealistic deterministic system of communication.' Rather, this randomness is an evolutionarily adaptive behavior. Deneubourg describes an experiment with two food sources near an ant nest: a rich food source far from the nest, and an inferior source close to the nest. Initially, the ants discover the inferior food source and form a robust trail to that source. But some ants wander off the trail. These 'lost ants' discover the richer source and form a trail to it. Since an ant's pheromone emissions are related to the richness of the food source, the trail to the richer source becomes stronger than the original trail. Eventually, most ants shift to the richer source. So the randomness of the ants provides a way for the colony to explore multiple food sources in parallel. While positive feedback encourages exploitation of particular sources, randomness encourages exploration of multiple sources." Resnick, 138.

81 "Typical teenagers," I: More on the ants' adolescent behavior: "Older colonies were more consistent from week-to-week than younger ones. I performed the same experiments, week after week, with several groups of older colonies and with several different groups of younger colonies. Week after week, each group of older colonies responded much as the other groups did, but week after week, each group of younger colonies had a different response. Curiously, in any given week, the variation among younger colonies was no greater than the variation among older colonies. The differences were only in week-to-week comparisons. Apparently younger colonies are more susceptible than older ones to changes in weather or to the amount of food available. In one week, a given perturbation pushes the young colonies in one direction; in another week, it pushes them in another." Gordon, 133.

81 "And the other": "Younger colonies were more willing than older ones to put up with neighbors in order to get food. In all colony pairs, regardless of colony age, both colonies foraged toward the seed bait when I put out seeds. In both age classes, the colonies foraged toward the bait even after it was gone. But younger colonies kept up longer. Younger colonies continued to forage toward the bait and to fight with each other for up to six days after the bait was gone. The older pairs gave up the conflict sooner. Once the food was gone, they were more likely to direct their foraging efforts elsewhere. For the younger colonies, a site that had offered abundant food for a few days before was still worth fighting over." Ibid., 51.

82 How does the: Gordon's theory is that the life cycle of the ant colony derives from population changes: "Since workers live only a year, the colony must re-create itself each year. The 4,000 ants of a 3-year-old colony have to feed 6,000 larvae to make the 4-year-old colony. For a colony 5 years or older, already at its adult size, the situation is different. For every new ant to feed, there is already an ant there to help feed it. The colony must produce 10,000 workers each year to maintain its size of 10,000 workers—but it has 10,000 workers to collect and process the food necessary to do this. So the demand for food, per forager, may be greater in the smaller, quickly growing colony. This might make foragers of a small, quickly growing colony more prone to engage in conflict over food than those of a larger one." Ibid., 83.

82 "The ancestors of": Ridley, 232.

83 And yet somehow: The argument extends down to the atomic level as well. "We can argue that consciousness and identity are not a function of the specific particles at all, because our own particles are constantly changing. On a cellular basis, we change most of our cells (although not our brain cells) over a period of several years. On an atomic level, the change is much faster than that, and does include our brain cells. We are not at all permanent collections of particles. It is the patterns of matter and energy that are semipermanent (that is, changing only gradually), but our actual material content is changing constantly, and very quickly." Kurzweil, 54.

85 The runaway power: "Exponential growth puts great power in the hands of naturally selected genes. It means that a tiny adjustment to a detail of embryonic growth control can have the most dramatic effect on the outcome. A mutation that tells a particular sub-lineage of cells to go on dividing just one more time—say go on for twenty-five cell generations instead of twenty-four—can in principle have the effect of doubling the size of a particular bit of the body. The same trick, of changing numbers of cell generations, or rates of cell division, can be used by genes during embryology to change the shape of a bit of the body. . . . In a way, the

remarkable thing is that cell lineages stop dividing when they are supposed to, in such a way that all our bits are well proportioned relative to one another." Dawkins, 1996, 293.

86 Cells rely heavily: ". . . within a time frame, a cell's competence depends upon its location, its previous history of locations, and the proximity of its neighbors in a collective. There is no indication that a cell's exact position in this collective is critical, but its fate can be determined by how many cells of similar history are located close by. A cell's fate thus depends upon its competence and upon its neighborhood." Edelman, 1988, 22.

86 Since every cell: Ridley, 175.

87 The words seem: Many brain researchers also draw upon the language of neighborhoods to describe how the brain develops. "Imagine now this epigenetic drama in which sheets of nerve cells in the developing brain form a neighborhood. Neighbors in that neighborhood exchange signals as they are linked by CAMs and CIMs. They send processes out in a profuse fashion, sometimes bunched together in bundles called fascicles. When they reach other neighborhoods and sheets they stimulate target cells." Edelman, 1992, 64.

88 Because each cell: Twenty years before Wright released SimCity, Thomas Schelling sketched out its basic principles in a decidedly low-tech game-theory experiment: "Get a roll of pennies, a roll of dimes, a ruled sheet of paper divided into one-inch squares, preferably at least the size of a checkerboard (sixty-four squares in eight rows and eight columns) and find some device for selecting squares at random. We place dimes and pennies on some of the squares, and suppose them to represent the members of two homogeneous groups—men and women, blacks and whites, French-speaking and English-speaking, officers and enlisted men, students and faculty, surfers and swimmers, the well-dressed and the poorly dressed, or any other dichotomy that is exhaustive and recognizable. We can spread them at random or put them in contrived patterns. We can use equal numbers of dimes and pennies or let one be a minority. And we can stipulate various rules for individual decisions." Schelling, 147.

89 "They differ from": Jacobs, 1961, 30.

89 "polycentric, plum-pudding": Similarly complex pattern formation has been observed in our friends the slime molds, as in the oft-observed concentric ring waves that ripple through the slime mold community at certain points. For years, researchers assumed that the cells at the center of the ring were responsible for the patterns, but new models suggest otherwise. As Garfinkel puts it: "Of course a set of concentric rings has a geometric center. But is the geometric center the *cause* of the propagation. Not necessarily." Garfinkel, 198.

89 Building on the: "A moderate urge to avoid small-minority status may cause a nearly integrated pattern to unravel, and highly segregated neighborhoods to form. Even a deliberately arranged viable pattern, as in Figure 3, when buffeted by a little random motion, proves unstable and gives way to the separate neighborhoods of Figures 5 through 8. These then prove to be fairly immune to continued random turnover. For those who deplore segregation, however, and especially for those who deplore more segregation than people were seeking when they collectively segregated themselves, there may be a note of hope. The underlying motivation can be far less extreme than the observable patterns of separation." Schelling, 154.

90 In any model: Krugman, 24–25.

91 "You'd be surprised": You know who you are.

94 And safety is: The emphasis on local knowledge is also key to the way certain "organic" cities develop over time. "Organic planning does not begin with a preconceived goal: it moves from need to need, from opportunity to opportunity, in a series of adaptations that themselves become increasingly coherent and purposeful, so that they generate a complex, final design, hardly less unified than a pre-formed geometric pattern. Towns like Siena illustrate this process to perfection. Though the last stage in such a process is not clearly present at the beginning, as it is in a more rational, non-historic order, this does not mean that rational considerations and deliberate fore-thought have not governed every feature of the plan, or that a deliberately unified and integrated design may not result." Mumford, 1961, 302.

95 Marshall Berman wrote: Berman, 347.

96 City life depends: Computer models of slime mold behavior also showcase the importance of random "swerves": "The StarLogo slime-mold project provides another example of randomness in the service of exploration. If the program had no randomness, slime-mold cells would rarely leave their clusters. The program would lose its dynamic and organic quality. The screen would become filled with lots of little clusters with little or no interchange of cells between clusters. The randomness in the program makes it more likely for cells to break free of their clusters. As a result, small clusters become less stable: when a small cluster loses one of its cells, the whole cluster is likely to break apart. Small clusters either grow or break apart. The result is fewer, larger clusters, with more cells moving from cluster to cluster. If the goal is for the slime-mold cells to aggregate into large clusters (as is the case with real slime-mold), then randomness plays a very useful role. Resnick, 139.

96 At sixty-five miles: Jacobs sees the deterioration of car-centric cities as a kind of feedback effect: "Erosion of cities by automobiles is thus an

example of what is known as 'positive feedback.' In cases of positive feedback an action produces a reaction which in turn intensifies the condition responsible for the first action. This intensifies the need for repeating the first action, which in turn intensifies the reaction, and so on, ad infinitum. It is something like the grip of a habit-forming addiction." Jacobs, 1961, 350.

98 world of metaphor: "[Humans] do resemble, in their most compulsively social behavior, ants at a distance. It is, however, quite bad form in biological circles to put it the other way 'round, to imply that the operation of insect societies has any relation at all to human affairs. The writers of books on insect behavior generally take great pains, in their prefaces, to caution that insects are like creatures from another planet. . . . They are more like crazy little machines, and we violate science when we try to read human meanings in their arrangements. It is hard for a bystander not to do so. Ants are so much like human beings as to be an embarrassment." Thomas, 11.

99 And that macrodevelopment: The growth of large social structures is sometimes best observed from the long view of outer space, where human development patterns uncannily resemble organic ones: "On larger scales, only occasionally does the work of our energetic species show up: a bridge, a wall, a dam, or a highway. These are typically less than fully three-dimensional. They seem long ribbons when occasionally they are caught in aerial views. Only in their collectivity do we see human artifacts that occupy large surface areas (still not three-dimensional) in the ten-to-hundred-kilometer range, sometimes even beyond. These are the cultivated plains and terraces, the irrigated lands, the clearings of the ancient forest, the great cities and their environs. Theirs is a history of growth more than one of design. For the rest of life, too, we find a similar display. Blades of grass are small, but grasslands and savannahs, like the dark forests north and south, extend over whole regions, easily up to a thousand kilometers across." Morrison, Eames, 2.

99 Those of us: "Tens of millions of people making billions of decisions every week about what to buy and what to sell and where to work and how much to save and how much to borrow and what orders to fill and what stocks to accumulate and where to move and what schools to go to and what jobs to take and where to build the supermarkets and movie theaters and electric power stations, when to invest in buildings above ground and mine shafts underground and fleets of trucks and ships and aircraft—if you are in a mood to be amazed, it can amaze you that the system works at all. Amazement needn't be admiration: once you understand the system you may think there are better ones, or better ways to

make this system work. I am only inviting you to reflect that whether this system works well or ill, in most countries and especially the countries with comparatively undirected economic systems, the system works the way ant colonies work." Schelling, 21.

102 If the engine: "In the medieval town these powers, the spiritual and the temporal, with their vocational orders, the warrior, the merchant, the priest, the monk, the bard, the scholar, the craftsman and the tradesman, achieved something like an equilibrium. That balance remained delicate and uncertain; but the effort to maintain it was constant and the effect real, because each social component was weighted, each duly represented. Until the close of the Middle Ages—this indeed is one of the signs of the close—no one element was strong enough to establish permanently its own command over all the others. As a result, both physically and politically, the medieval city, though it recapitulated many of the features of the earliest urban order, was in some respects an original creation." Mumford, 1961, 252.

102 Merchants who were: Hibbert, 1993, 102.

103 Those antibodies function: "The immune system is a somatic selective system consisting of molecules, cells, and specialized organs. As a system, it is capable of telling the difference between self and nonself at the molecular level. For example, it is responsible for distinguishing between and responding to the chemical characteristics of viral and bacterial invaders (nonself), invaders that would otherwise overwhelm the collections of cellular systems in an individual organism (self). This response involves molecular recognition with an exquisite degree of specificity. An appropriately stimulated immune system can tell the difference between two large foreign protein molecules composed of thousands of carbon atoms that differ by only a few degrees in the tilt of a single carbon chain. It can tell these molecules apart from all other molecules and retain the ability to do so once it has initially developed that ability. It has a 'memory.'" Edelman, 1992, 75.

103 What's equally amazing: "The immune selective system has some intriguing properties. First, there is more than one way to recognize successfully any particular shape. Second, no two individuals do it exactly the same way; that is, no two individuals have identical antibodies. Third, the system has a kind of cellular memory. After the presentation of an antigen to a set of lymphocytes that bind it, some will divide only a few times, while the rest go on irreversibly to produce antibody specific for that antigen and die. Because some of the cells have divided but not all the way to the antibody-making end, they constitute a larger group of cells in the total population of cells than were originally present. This larger group

can respond at a later time in an accelerated fashion to the same antigen. As I mentioned before, the system therefore exhibits a form of memory at the cellular level." Ibid., 78.

105 The world convulses: "Note one more feature: the neighborhood unit and the functional precinct. In a sense, the medieval city was a congeries of little cities, each with a certain degree of autonomy and self-sufficiency, each formed so naturally out of common needs and purposes that it only enriched and supplemented the whole. The division of the town into quarters, each with its church or churches, often with a local provision market, always with its own local water supply, a well or a fountain, was a characteristic feature; but as the town grew, the quarters became sixths, or even smaller fractions of the whole, without dissolving into the mass. . . . This integration into primary residential units, composed of families and neighbors, was complemented by another kind of division, into precincts, based on vocation and interest: thus both primary and secondary groups, both Geminschaft and Gesellschaft, took on the same urban pattern. In Regensburg, as early as the eleventh century, the town was divided into a clerical precinct, a royal precinct, and a merchant's precinct, corresponding to the chief vocations, while craftsmen and peasants must have occupied the rest of the town." Mumford, 1961, 310.

105 That pattern in: "It is a form of collective memory—closer in a way to the sense in which the body develops memories than the way in which the conscious mind does. I am using the word *memory* here in a more inclusive fashion than usual. Memory is a process that emerged only when life and evolution occurred and gave rise to the systems described by the sciences of recognition. As I am using the term *memory,* it describes aspects of heredity, immune responses, reflex learning, true learning following perceptual categorization, and the various forms of consciousness. . . . Memory is an essential property of biologically adaptive systems." Edelman, 1992, 203–4.

107 "From its origins": Mumford, 1961, 30.

107 That clustering becomes: "[These economies] come from the fact that the firm can find in the large city all manner of client, services and suppliers, and employees no matter how specialized its product; this, in turn, promotes increased specialization. Surprisingly, however, economies of agglomeration encourage firms of the same line to locate close to one another, which is why names such as Harley, Fleet, and Lombard Streets and Savile Row—to stick to London—call to mind professions rather than place. Besides the non-negligible profit and pleasure of shop-talk, all can share access to services that none could support alone. . . . A key point about economies of agglomeration is that small businesses depend

on them more than do large ones. The latter can internalize these 'external economies' by providing their own services and gain locational freedom as a result. . . . The relationship between large cities and small businesses is a symbiotic one beneficial to both. The reason is that small firms are the major carriers of innovation, including creative adaptation to change. This was even more true in the days before scientific research contributed much to the new technology." From Hohenberg and Lees, *The Making of Urban Europe,* quoted in De Landa, 1997, 85–86.

108 But cities have: "Though the great city is the best organ of memory man has yet created, it is also—until it becomes too cluttered and disorganized—the best agent for discrimination and comparative evaluation, not merely because it spreads out so many goods for choosing, but because it likewise creates minds of large range, capable of coping with them. Yes: inclusiveness and large numbers are often necessary; but large numbers are not enough. Florence, with some four hundred thousand inhabitants, performs more of the functions of the metropolis than many other cities with ten times that number." Mumford, 1961, 562.

108 By some accounts: "The most ancient urban remains now known, except Jericho, date from this period. This constituted a singular technological expansion of human power whose only parallel is the change that has taken place in our own time." Ibid., 33.

108 The neighborhood system: Decades before the first graphical interface was designed, Wiener connected the problems of communal information and software interface, gesturing to Vannevar Bush's visionary essay on the Memex: "On the other hand, the human organism contains vastly more information, in all probability, than does any one of its cells. There is thus no necessary relation in either direction between the amount of racial or tribal or community information and the amount of information available to the individual. . . . As in the case of the individual, not all the information which is available to the race at one time is accessible without special effort. There is a well-known tendency of libraries to become clogged by their own volume; of the sciences to develop such a degree of specialization that the expert is often illiterate outside his own minute specialty. Dr. Vannevar Bush has suggested the use of mechanical aids for the searching through vast bodies of material." Wiener, 158.

109 The specialization of: "Early in the evolution of life-forms, specialized organs developed the ability to maintain internal states and respond differently to external stimuli. The trend ever since has been toward more complex and capable nervous systems with the ability to store extensive memories; recognize patterns in visual, auditory, and tactile stimuli; and engage in increasingly sophisticated levels of reasoning. The ability to

remember and to solve problems—computation—has constituted the cutting edge in the evolution of multicellular organisms." Those different skills—memory, pattern recognition, computation—all have parallels to the development of urban centers: in their clusters of shared information, their capacity to reflect and amplify patterns of human behavior, their mastery of complicated supply-and-demand problems. Kurzweil, 18.

109 They are the: The potential advantages of high population densities are evident even in primitive societies. "How to keep these costs low if your communications and transportation technologies are primitive? One way is to stay near your customers and suppliers. In other words: live in a society with high population density. This may be the key to the wealth of the American Northwest: not natural abundance per se—an abundance quickly diluted by thick population, anyway—but rather the thick population that does the diluting. Back before communications and transportation were sufficiently high tech to catalyze markets, the stimulus came instead from a habitat that would tolerate large, close populations." Wright, 47.

110 As the physicist: "Iberall was perhaps the first to view the major transitions in early human history (the transitions from hunter-gatherer to agriculturalist, and from agriculturalist to city dweller) not as a linear advance up the ladder of progress but as the crossing of nonlinear critical thresholds (bifurcations). More specifically, much as a given chemical compound (water, for example) may exist in several distinct states (solid, liquid, or gas) and may switch from stable state to stable state at critical points in the intensity of temperature (called phase transitions), so a human society may be seen as a 'material' capable of undergoing these changes of state as it reaches critical mass in terms of density of settlement, amount of energy consumed, or even intensity of interaction.

"Iberall invites us to view early hunter-gatherer bands as gas particles, in the sense that they lived apart from each other and therefore interacted rarely and unsystematically. (Based on the ethnographic evidence that bands typically lived about seventy miles apart and assuming that humans can walk about twenty-five miles a day, he calculates that any two bands were separated by more than a day's distance from one another.) When humans first began to cultivate cereals and the interaction between human beings and plants created sedentary communities, humanity liquefied or condensed into groups whose interactions were now more frequent although still loosely regulated. Finally, when a few of these communities intensified agricultural production to the point where surpluses could be harvested, stored, and redistributed (for the first time allowing a division of labor between producers and consumers of food), humanity acquired a crystal state, in the sense that central governments now imposed a symmetrical grid of laws and regulations on the urban populations." De Landa, 1997, 15.

111 Cities aren't ideas: "Is the city a natural habitation, like a snail's shell, or a deliberate human artifact, a specific invention that came into existence at one or more places under the influence of urban ideological convictions and economic pressures? An aboriginal predisposition toward social life, even toward group settlement, may well characterize the human species; but could such a general tendency make man everywhere produce the city as inevitably as a spider produces her web? Could the same dispositions that gave the camp or the hamlet a planetary distribution likewise account for such a many-faceted cultural complex as the city?" Mumford, 1961, 185.

111 A linear increase: "One of the first cultural evolutionists to emphasize energy technologies was Leslie White. Indeed, some mark the mid-twentieth-century renaissance of interest in cultural evolutionism to the publication in 1943 of his paper 'Energy and the Evolution of Culture.' White was also among the cultural evolutionists who more or less ignored information technologies. In a way, this is not surprising. His landmark paper came out a year before Schrödinger's book appeared, a decade before DNA was discerned, and decades before science had truly grasped the pervasive role of biological information in upholding the integrity of organisms. Thus his attempts to import insights from biology into the social sciences met with limited success. He observed that the question 'What holds systems together?' is 'as fundamental to sociology as it is to biology,' but he couldn't carry the analysis further." Wright, 249.

112 As the historian: Quoted in De Landa, 1997, 29.

112 As Mumford writes: Mumford, 1961, 258.

112 The result is: Of course, changes in energy flows produced more than just the urban explosion of the late Middle Ages. "Indeed, urban morphogenesis has depended, from its ancient beginnings in the Fertile Crescent, on intensification of the consumption of nonhuman energy. The anthropologist Richard Newbold Adams, who sees social evolution as just another form that the self-organization of energy may take, has pointed out that the first such intensification was the cultivation of cereals. Since plants, via photosynthesis, simply convert solar energy into sugars, cultivation increased the amount of solar energy that traversed human societies. When food production was further intensified, humanity crossed the bifurcation that gave rise to urban structures. The elites that ruled those early cities in turn made other intensifications possible—by developing large irrigation systems, for example—and urban centers

mutated into their imperial form. It is important to emphasize, however, that cereal cultivation was only one of several possible ways of intensifying energy flow. As several anthropologists have pointed out, the emergence of cities may have followed alternative routes to intensification, as when the emergence of urban life in Peru fed off a reservoir of fish. What matters is not agriculture per se, but the great increase in the flow of matter-energy through society, as well as the transformations in urban form that this intense flow makes possible." De Landa, 1997, 28.

112 We sometimes talk: There's a more delicate way to express the same idea, which is that those early urbanites were collectively engaged in shortening their nutrient cycles. "By shortening food chains, human populations acquired control over nutrient cycles. For instance, cattle and certain crops went hand-in-hand: the manure of the cattle, which were raised on cereals, could be plugged back into the system as fertilizer, closing the nutrient cycle. In itself, this tightening of the cycles was good. Indeed, ecosystems spontaneously shorten their nutrient cycles as they complexify. A highly complex system such as a rain forest runs its nutrients so tightly, via elaborate microflora and microfauna in the tree roots, that the soil is largely deprived of nutrients. This is one reason why the destruction of rain forests is so wasteful: the soil left behind is largely sterile." Ibid., 122.

112 "This acceleration in": Ibid., 29.

114 Some critics, such: Wright sees "group brains" even in low-tech societies, even using the term as an alternative to "invisible hands." "Hands aren't very cerebral, after all; guiding any invisible hand there must be an 'invisible brain.' Its neurons are people. The more neurons there are in regular and easy contact, the better the brain works—the more finely it can divide economic labor, the more diverse the resulting products. And, not incidentally, the more rapidly technological *innovations* take shape and spread. As economists who espouse 'new growth theory' have stressed, it takes only one person to invent something that the whole group can then adopt (since information is a 'non-rival' good). So the more possible inventors—that is, the larger the group—the higher its collective rate of innovation. All told, then, the Northwest Coast Indians outproduced and outinvented the Shoshone not because they had better brains (the sort of conclusion Franz Boas worried about) but because they *were* a better brain." Wright, 48.

Turing relied on a similarly abstract notion of what a brain is in developing the modern computer: "To understand the Turing model of 'the brain,' it was crucial to see that it regarded physics and chemistry, including all the arguments about quantum mechanics to which Eddington had appealed, as essentially irrelevant. In his view, the physics and chemistry were relevant only in as much as they sustain the medium for the embodiment of discrete 'states,' 'reading' and 'writing.' Only the logical pattern of these 'states' could really matter. The claim was that whatever a brain did, it did by virtue of its structure as a logical system, and not because it was inside a person's head, or because it was spongy tissue made up of a particular kind of biological cell formation. And if this were so, then its logical structure could just as well be represented in some other medium, embodied by some other physical machinery. It was a materialist view of mind, but one that did not confuse logical patterns and relations with physical substances and things, as so often people did." Hodges, 291.

116 "Not as crazy": Wright, 302.

118 "But the Internet": From a *Slate* book club, February 1, 2000.

122 "So the question": Interviews with Brewster Kahle, conducted October 2000 and July 1998.

125 Decades ago, in: Wiener, 35.

126 Our brains got: Once again, the information-processing skills of ant colonies are instructive here: ". . . it is tempting to speculate about the generality of interaction patterns as a source of information in natural systems. What I like about the idea that an ant's task decision is based on its interaction rate is that the pattern of interaction, not a signal in the interaction itself, produces the effect. Ants do not tell each other what to do by transferring messages. The signal is in the pattern of contact. Such a process might operate in brains, immune systems, or any place where the rate of flow of a certain type of unit, or the activity level of a certain type of unit, is related to the need for a change in the rate of flow. Interaction rate is the local translation of a characteristic of the whole system, rate of flow or activity, and each unit's reaction to this local cue contributes to a predictable response by the whole system." Gordon, 169.

127 The human mind: Kurzweil, 77.

127 But unlike most: Ibid., 103.

127 "We then use": Ibid., 77.

127 Certainly the evidence: Turing from an interview in the late forties: "This is only a foretaste of what is to come, and only the shadow of what is going to be. We have to have some experience with the machine before we really know its capabilities. It may take years before we settle down to the new possibilities, but I do not see why it should not enter any one of the fields normally covered by the human intellect, and eventually compete on equal terms." Hodges, 406.

128 What drives each: Returning us to the roots of computing and Turing's breaking the code of the Enigma device: "The task of the analyst, corre-

spondingly, was to determine this ring-setting which was common to all the traffic of the network. . . . As with the older method, the fingerprint depended upon looking at the entire traffic, and in exploiting the element of repetition in the last six of the nine indicator letters. Without a common ground-setting, there was no fixed correspondence between first and fourth, second and fifth, third and sixth letter, to analyze." Ibid., 173.

130 "She claims she": Rosenstiel, 55–65.

133 But beneath all: This is true also for the interaction between the brain and the rest of the body. "Nervous system behavior is to some extent self-generated in loops; brain activity leads to movement, which leads to further sensation and perception and still further movement. The layers and the loops between them are the most intricate of any object we know, and they are dynamic; they continually change." Edelman, 1992, 29.

134 A given circuit: "What is learning? What changes occur to nerve cells when the brain (or the abdominal ganglion) acquires a new habit or a change in its behavior? The central nervous system consists of lots of nerve cells, down each of which electrical signals travel; and synapses, which are junctions between nerve cells. When an electrical nerve signal reaches a synapse, it must transfer to chemical agent, like a train passenger catching a ferry across a sea channel, before resuming its electrical journey. Kandel's attention quickly focused on these synapses between neurons. Learning seems to be a change in their properties. Thus when a sea slug habituates to a false alarm, the synapse between the receiving, sensory neuron and the neuron that moves the gill is somehow weakened. Conversely, when the sea slug is sensitized to the stimulus, the synapse is strengthened." Ridley, 223.

134 If each neuron: Edelman has a far more precise variation on this theme, which he calls "reentry." "To explain how categorization may occur, we can use the workings of what I have called a 'classification couple' in the brain. This is a minimal unit consisting of two functionally different maps made up of neuronal groups and connected by reentry. Each map independently receives signals from other brain maps or from the world (in this example the signals come from the world). Within a certain time period, reentrant signaling strongly connects certain active combinations of neuronal groups in one map to different combinations in the other map." Edelman, 1992, 87.

136 The Flowers episode: Rosenstiel, 63.

137 The feedback loops: Jacobs wrote about this in *Death and Life* as the tendency for successful cities to destroy themselves: "These forces, in the form that they work for ill, are: the tendency for outstandingly successful

diversity in cities to destroy itself; the tendency for massive single elements in cities (many of which are necessary and otherwise desirable) to cast a deadening influence; the tendency for population instability to counter the growth of diversity; and the tendency for both public and private money either to glut or to starve development and change." Jacobs, 1961, 242.

137 In the Flowers: As we saw in the last chapter, positive feedback is also an important tool for understanding social or technological revolutions: "These meshworks of mutually supporting innovations (coal-iron-steam-cotton) are well-known to historians of technology. They existed long before the nineteenth century (e.g., the interlocking web formed by the horseshoe, the horse harness, and triennial rotation which was behind the agricultural intensification at the turn of the millennium), and they occurred afterward, as in the meshwork of oil, electricity, steel, and synthetic materials that contributed to the second industrial revolution. Nonetheless, as important as they were, autocatalytic loops of technologies were not complex enough to create a self-sustained industrial take-off. Before the 1800s, as we noted, these intensifications often led back to depletions of resources and diminishing returns. Negative feedback eventually checked the turbulent growth generated by positive feedback." De Landa, 1997, 77.

139 But the new: As usual, Jane Jacobs was quick to adapt these new ideas to her understanding of the city: "The analogy that comes to mind is faulty feedback. The conception of electronic feedback has become familiar with the development of computers and automated machinery, where one of the end products of an act or series of acts by the machine is a signal which modifies and guides the next act. A similar feedback process, regulated chemically rather than electronically, is now believed to modify some of the behavior of cells. A report in the *New York Times* explains it thus:

> "'The presence of an end product in the milieu of a cell causes the machinery that produces the end product to slow down or to stop. This form of cell behavior Dr. [Van R.] Potter [of the University of Wisconsin Medical School] characterized as "intelligent." In contrast, a cell that has changed or mutated behaves like an "idiot" in that it continues without feedback regulation to produce even materials that it does not require.'
>
> "I think that last sentence is a fair description of the behavior of city localities where the success of diversity destroys itself.
>
> "Suppose we think of successful city areas, for all their extraordinary and intricate economic and social order, as faulty in this

fashion. In creating city success, we human beings have created marvels, but we left out feedback. What can we do with cities to make up for this omission?" Jacobs, 1961, 251–52.

140 As Wiener puts: Wiener, 7.

140 Wiener gave that: "In short, our inner economy must contain an assembly of thermostats, automatic hydrogen-ion-concentration controls, governors, and the like, which would be adequate for a great chemical plant. These are what we know collectively as our homeostatic mechanism." Ibid., 115.

140 Your body is: Ecosystems too abound with feedback systems. As one of the characters in Jane Jacob's latest book says, "Here's a pleasing example of this category—a positive loop in a California coastal redwood forest. Mature redwoods require enormous amounts of water, about twice as much, on average, as rainfall delivers to their habitats. . . . A coastal redwood lives to an age of about two thousand years; quite a demonstration of successful survival. Here's how their seemingly inadequate supply situation is overcome. With their find and luxuriant needles, the trees intercept fog and strip its moisture; in effect, they take water straight from clouds. During a dry but foggy night, each tall redwood douses the ground beneath it with as much water as if there had been a drenching rainstorm.This beneficent process works as a loop. The growth of the trees is fed in good part from the fog. Taller growth gives trees access to higher—hence additional—fog. Additional fog feeds still higher growth. And so on. Because of the height-fog loop, the trees themselves participate in keeping their environment stable. " Jacobs, 2000, 93.

142 If analyzing indirect: Dean.

142 Once you've reached: "There is already one technology that appears to generate at least one aspect of a spiritual experience. This experimental technology is called Brain Generated Music (BGM), pioneered by NeuroSonics, a small company in Baltimore, Maryland, of which I am a director. BGM is a brain-wave biofeedback system capable of evoking an experience called the Relaxation Response, which is associated with deep relaxation. The BGM user attaches three disposable leads to her head. A personal computer then monitors the user's brain waves to determine her unique alpha wavelength. Alpha waves, which are in the range of eight to thirteen cycles per second (cps), are associated with a deep meditative state, as compared to beta waves (in the range of thirteen to twenty-eight cps), which are associated with routine conscious thought. Music is then generated by the computer, according to an algorithm that transforms the user's own brain-wave signal. Kurzweil, 157.

143 "In connection with": Wiener, 158.

143 He would have: ". . . one of the directions of work which the realm of ideas of the Macy meetings has suggested concerns the importance of the notion and the technique of communication in the social system. It is certainly true that the social system is an organization like the individual, that it is bounded together by a system of communication, and that it has a dynamics in which circular processes of a feedback nature play an important part." Ibid., 24.

146 But the book: "The program he proposed in 1898 was to halt the growth of London and also repopulate the countryside, where villages were declining, by building a new kind of town—the Garden City, where the city poor might again live close to nature. So they might earn their livings, industry was to be set up in the Garden City, for while Howard was not planning cities, he was not planning dormitory suburbs either. His aim was the creation of self-sufficient small towns, really very nice towns if you were docile and had no plans of your own and did not mind spending your life among others who had no plans of their own. As in all Utopias, the right to have plans of any significance belonged only to the planners in charge." Jacobs, 1961, 17.

146 Better to build: "Ebenezer Howard's vision of the Garden City would seem almost feudal to us. He seems to have thought that members of the industrial working classes would stay neatly in their class, and even at the same job within their class; that agricultural workers would stay in agriculture; that businessmen (the enemy) would hardly exist as a significant force in his Utopia; and that planners could go about their good and lofty work, unhampered by rude nay-saying from the untrained." Ibid., 289.

147 His attachment to: "Howard's greatest contribution was less in recasting the physical form of the city than in developing the organic concepts that underlay this form; for though he was no biologist like Patrick Geddes, he nevertheless brought to the city the essential biological criteria of dynamic equilibrium and organic balance: balance as between city and country in a larger ecological pattern, and balance between the varied functions of the city: above all, balance through the positive control of growth in the limitation in area, number, and density of occupation, and the practice of reproduction (colonization) when the community was threatened by such an undue increase in size as would lead only to lapse of function." Mumford, 1961, 516.

147 "Any ecological association": Mumford, 1962, 148–77.

147 Like many debates: Wiener had made the same connection a decade before in *Cybernetics*: "It has been commented on by many writers, such as D'Arcy Thompson, that each form of organization has an upper limit of

size, beyond which it will not function. Thus the insect organization is limited by the length of tubing over which the spiracle method of bringing air by diffusion directly to the breathing tissues will function; a land animal cannot be so big that the legs or other portions of contact with the ground will be crushed by the weight; a tree is limited by the mechanism for transferring water and minerals from the roots to the leaves, and the products of photosynthesis from the leaves to the roots, and so on." Wiener, 150.

153 "The signal was": Posted on the Slashdot site: www.slashdot.org.

155 "As in the legal analogy": Technically, Slashdot moderators don't give each post a grade on the scale. Posts start out life at 0 or 1 (depending on whether their authors are registered users of the system.) Moderators can then "spend" a moderation point rating a post either up or down. A post that starts life at 1, and receives three positive points and one negative point would be at Level 3, because 1 plus 3 minus 1 equals 3.

156 "He was far": Jacobs, 2000, 154. A related idea is the pricing mechanism of market economies as an information-processing system, as described by the libertarian demigod Friedrich von Hayek. "Long before the fall of communism, Hayek identified its oft-overlooked weakness: not only did it fail to offer an incentive to work hard; it forced signals connecting supply and demand to travel a tortuous path that invited distortion." Wright, 199.

156 "Others were busy": Interview conducted with Rob Malda, April 2000.

164 A descendant of: The ways in which Resnick altered the Logo turtle model are instructive: "First, StarLogo has lots more turtles. Whereas commercial versions of Logo typically have only a few turtles, StarLogo has thousands of turtles. And StarLogo is designed as a massively parallel language—so all of the turtles can perform their actions at the same time, in parallel. . . . Second, StarLogo turtles have better 'senses.' The traditional Logo turtle was designed primarily as a 'drawing turtle,' for creating geometric shapes and exploring geometric ideas. But the StarLogo turtle is more of a 'behavioral turtle.' StarLogo turtles come equipped with 'senses.' They can detect (and distinguish) other turtles nearby, and they can 'sniff' scents in the world. . . . Third, StarLogo reifies the turtles' world. In traditional versions of Logo the turtles' world does not have many distinguishing features. The world is simply a place where the turtles draw with their pens. Each pixel of the world has a single piece of state information—its color. StarLogo attaches a much higher status to the turtles' world. The world is divided into small square sections called patches. (The term *patch* is borrowed from Pauline Hogeweg [1989].) The patches have many of the same capabilities as turtles—except that they cannot move." Resnick, 33–34.

164 "'Smelling' the green": Interviews with Resnick conducted in May 2000 and November 1999.

168 "even *Marvin Minsky*": Resnick, 119–20.

169 Then you press: Deborah Gordon observes a comparable phenomenon with her harvester ant colonies: "One lesson from the ants is that to understand a system like theirs, it is not sufficient to take the system apart. The behavior of each unit is not encapsulated inside that unit but comes from its connections with the rest of the system. To see how the components produce the response of the whole system, we have to track these connections in changing situations. You could dissect a brain into millions of separate nerve cells but would never find any dedicated to thinking about 'nature,' or 'ants,' or anything else; thoughts are made by the shifting pattern of interactions of neurons. Antibodies form in the immune system as a consequence of encounters with foreign cells. Ants are not born to do a certain task; an ant's function changes along with the conditions it encounters, including the activities of other ants." Gordon, 168.

172 Hillis's software was: Levy, 195–200.

173 "It may be": Hillis, 146.

173 In the short term: Ibid., 138.

178 "One of the": Interview conducted with Zimmerman, February 2000.

183 "The new sets": Interview conducted with Heywood, October 2000.

188 "The problem is": Interview conducted with Wright, October 2000.

193 "Can a selectional": Edelman, 1992, 190.

197 It's the chimp: De Waal, 49.

199 Rizzollati called these: Alison Motluk, "Read My Mind," *The New Scientist,* January 27, 2001.

199 They are mind: "Using a totally different test (the Smarties test), Perner, Frith, Leslie, and Leekham got the same basic result. In this test, the child is first shown a familiar Smarties container and is asked, 'What do you think is in here?' The child naturally replies, 'Smarties.' The child is then shown the tube actually contains pencils. Next the experimenter closes the tube, asks the child two belief questions. The first question is 'When I showed you this tube [before we opened it up], what did you think was in here?' The normal child, of course, correctly replies by referring to his earlier, now false, belief: 'Smarties.' The experimenter follows this up with: 'And when the next child comes in [who hasn't seen the tube], what will he think is inside here?' Again, the normal child correctly replies by referring to the other child's false belief: 'Smarties.' When Perner et al. gave this task to children with autism, they found that the majority of their subjects answered, 'Pencils,' to the two belief questions. That is, they answered by considering their own knowledge of what was

in the box rather than by referring to their own previous false belief or to someone else's current false belief. The robustness of this finding suggests that in autism there is a genuine inability to understand other people's different beliefs." Baron-Cohen, 70–71.

200 We're conscious of: Ibid., 130.

201 "An absence of": Dennett, 1991, 324.

202 Only when we: Ray Kurzweil refers to this as the "Consciousness Is Just a Machine Reflecting on Itself" School. " . . . consciousness is not exactly an illusion, but just another logical process. It is a process responding and reacting to itself. We can build that in a machine: just build a procedure that has a model of itself, and that examines and responds to its own methods. Allow the process to reflect on itself. There, now you have consciousness. It is a set of abilities that evolved because self-reflective ways of thinking are inherently more powerful." Kurweil, 58.

202 The great preponderance: Social interaction is deeply intertwined with brain chemistry: "The higher your self-esteem and social rank relative to those around you, the higher your serotonin level is. Experiments with monkeys reveal that it is the social behavior that comes first. Serotonin is richly present in dominant monkeys and much more dilute in the brains of subordinates. Cause or effect? Almost everybody assumed the chemical was at least partly the cause: it just stands to reason that the dominant behavior results from the chemical, not vice versa. It turns out to be the reverse: serotonin levels respond to the monkey's perception of its own position in the hierarchy, not vice versa." Ridley, 170.

202 Among the apes: Baron-Cohen, 15.

202 Orangutans live mostly: Diamond, 1997.

202 Pleistocene-era experts: "That there was a massive neurocognitive evolution during the Pleistocene epoch is beyond any doubt. The brain has increased threefold in size in the 3 million years since *Australopithecus afarensis* evolved, going from around 400 cubic centimeters to its current size of about 1350 cubic centimeters.

The increase in brain size is likely to have had many causes, but one key factor upon which many theorists agree is the need for greater social intelligence shorthand for the ability to process information about the behavior of others and to react adaptively to their behavior. It is likely that there was a need for greater social intelligence because the vast majority of nonhuman primate animals are social animals, living in groups that range from as few as two individuals to as many as two hundred." Baron-Cohen, 13–14.

203 We don't know: ". . . the network of the brain is created by cellular movement during development and by the extension and connection of

increasing numbers of neurons. The brain is an example of a self-organizing system. And examination of this system during its development and of its most microscopic ramifications after development indicates that precise point-to-point wiring (like that in an electronic device) cannot occur. The variation is too great." Edelman, 1992, 25.

204 No individual neuron: "'When we hear it said that wireless valves think,' Jefferson said, 'we may despair of language.' But no cybernetician had said the *valves* thought, no more than anyone would say that the nerve cells thought. Here lay the confusion. It was the system as a whole that 'thought,' in Alan Turing's view, and it was its logical structure, not its particular physical embodiment, that made this possible." Hodges, 405.

204 By following the: How "natural" these solutions are remains an open question. "Is the city a natural habitation, like a snail's shell, or a deliberate human artifact, a specific invention that came into existence at one or more places under the influence of urban ideological convictions and economic pressures? An aboriginal predisposition toward social life, even toward group settlement, may well characterize the human species; but could such a general tendency make man everywhere produce the city as inevitably as a spider produces her web? Could the same dispositions that gave the camp or the hamlet a planetary distribution likewise account for such a many-faceted cultural complex as the city?" Mumford, 1961, 90.

205 A community of: These bottom-up solutions are not entirely unopposed. You can make the argument that the real battle of the next ten years on the Web is the battle between hierarchical forces (AOL Time Warner, the Chinese government) and the decentralized forces described in this book. As De Landa observes, "Although antimarket institutions had an early presence in the computer meshwork, today they are set to invade the Internet with unprecedented force. It is possible that the meshworks that have already accumulated within the Internet will prove resilient enough to survive the attack and continue to flourish. It is also possible in the next decades that hierarchies will instead accumulate, perhaps even changing the network back into a one-to-many system of information delivery. The outcome of this struggle has certainly not been settled." De Landa, 1997, 254.

209 "In other words": Interview conducted with Wright in October 2000.

222 Marketplaces—even those: The hierarchical nature of the modern corporation is not a new development. The historian Fernand Braudel goes so far as to describe capitalist structures as intrinsically top-down ones, opposed by lower-level, decentralized market forces. "There is a dialectic still very much alive between capitalism on the one hand, and its antithesis, the 'noncapitalism' of the lower level on the other. . . . [This] lower

level, not being paralyzed by the size of its plant or organization, is the one readiest to adapt; it is the seedbed of inspiration, improvisation, and even innovation, although its most brilliant discoveries sooner or later fall into the hands of the holders of capital. It was not the capitalists who brought about the first cotton revolution; all the new ideas came from enterprising small businesses." Quoted in De Landa, 1997, 46.

223 Those qualities make: "Many organizations nowadays are consciously trying to figure out how they can use self-organizing principles without becoming either disintegrated or inert—in short, as avatars of fruitful complexity. Ecotrust lists these three requirements: (a) autonomous agents able to make independent decisions within a framework of relatively simple rules; (b) moderately dense network and web connections among the agents—that is, the organization's parts; and (c) vigorous experimentation by agents, disciplined by responding to feedback on results." Jacobs, 2000, 177.

224 The Australian software: *Emergence,* 46.

226 Klein writes, "What": Klein.

226 By any measure: As always, building emergent systems doesn't guarantee that they will turn out to be better than the old systems. You need to get the variables right. "The mere presence of an emergent meshwork does not in itself mean that we have given a segment of society a less oppressive structure. The nature of the result will depend on the character of the heterogeneous elements meshed together, as we observed of communities on the Internet: they are undoubtedly more destratified than those subjected to massification by one-to-many media, but since everyone of all political stripes—even fascists—can benefit from this destratification, the mere existence of a computer meshwork is no guarantee that a better world will develop there." De Landa, 1997, 272.

228 Dorgo's secret: Bonabeau and Thiraulaz, 73.

BIBLIOGRAPHY

Alexander, Christopher, and Sara Ishikawa et al. *A Pattern Language: Towns, Buildings, Construction.* New York: Oxford University Press, 1977.

Axlerod, Robert. *The Evolution of Cooperation.* New York: Basic Books, 1984.

Bak, Per. *How Nature Works: The Science of Self-Organized Criticality.* New York: Springer-Verlag, 1996.

Ball, Philip. *The Self-Made Tapestry: Pattern Formation in Nature.* New York, Oxford, and Tokyo: Oxford University Press, 1999.

Baron-Cohen, Simon. *Mindblindness: An Essay on Autism and Theory of Mind.* Cambridge, Mass., and London: MIT Press, 1999.

———, ed. *The Maladapted Mind: Classic Readings in Evolutionary Psychopathology.* East Sussex, UK: Psychology Press, 1997.

Becker, Konrad, and Miss M. "An Interview with Manuel De Landa." Online posting. www.t0.or.at/delanda/intdeladna.htm. May 2000.

Benjamin, Walter. *The Arcades Project.* Cambridge, Mass., and London: Harvard University Press, 1999.

Bernardini, Wesley. "Transitions in Social Organization: A Predictive Model from Southwestern Archaeology." *Journal of Anthropological Archaeology* 15 (1996): 372–402.

Berners-Lee, Tim. *Weaving the Web: The Original Design and Ultimate Destiny of the World Wide Web by Its Inventor.* New York: HarperCollins, 1999.

Bonabeau, Eric, and Guy Thiraulaz. "Swarm Smarts." *Scientific American,* March 2000, 73–79.

Borsook, Eve. *The Companion Guide to Florence.* Englewood Cliffs, N.J.: Prentice-Hall, 1973.

Brand, Stewart. *How Buildings Learn: What Happens After They're Built.* New York and London: Penguin Books, 1994.

Braudel, Fernand. *A History of Civilizations.* New York and London: Penguin Books, 1993.

———. *The Perspective of the World.* Vol. 3 of *Civilization and Capitalism, 15th–18th Century.* Berkeley and Los Angeles: University of California Press, 1992.

———. *The Wheels of Commerce.* Vol. 2 of *Civilization and Capitalism, 15th–18th Century.* Berkeley and Los Angeles: University of California Press, 1992.

Briggs, Asa. *Victorian Cities.* Berkeley and Los Angeles: University of California Press, 1963, 1970.

Brosterman, Norman. *Inventing Kindergarten.* New York: Harry N. Abrams, 1997.

Buck-Morss, Susan. *Dialectics of Seeing.* Cambridge, Mass.: MIT Press, 1990.

Burrows, Edwin G., and Mike Wallace. *Gotham: A History of New York City to 1895.* New York and Oxford: Oxford University Press, 1999.

Calvin, William. *The Cerebral Code: Thinking a Thought in the Mosaics of the Mind.* Cambridge, Mass., and London: MIT Press, 1996.

Clark, Andy. *Being There: Putting Brain, Body and World Together Again.* London and Cambridge, Mass.: MIT Press, 1997.

Coleman, Henry J., Jr. "What Enables Self-Organizing Behavior in Business." *Emergence* 1, no. 1 (1999): 33–48.

Collins, George R., and Christiane Crasemann Collins. *Camillo Sitte: The Birth of Modern City Planning.* New York: Rizzoli, 1986.

Connolly, Peter, and Hazel Dodge. *The Ancient City: Life in Classical Athens and Rome.* New York and Oxford: Oxford University Press, 1998.

Corbusier, Le. *Towards a New Architecture.* New York: Dover Publications, 1986.

Davis, Mike. *City of Quartz: Excavating the Future in Los Angeles.* New York: Vintage Books, 1992.

Dawkins, Richard. *Climbing Mount Improbable.* New York and London: W. W. Norton, 1996.

———. *The Extended Phenotype: The Long Reach of the Gene.* New York and Oxford: Oxford University Press, 1982.

———. *Unweaving the Rainbow: Science, Delusion and the Appetite for Wonder.* London: Penguin Press, 1998.

Dean, Katie. "Attention Kids: Play This Game." *Wired News.* December 19, 2000.

Dehaene, Stanislas, Michel Kerszberg, and Jean-Pierre Changeux. "A neuronal model of a global workspace in effortful cognitive tasks." *Proceedings of the National Academy of Sciences of the United States of America* 95 (1998): 14529–34.

De Landa, Manuel. *A Thousand Years of Nonlinear History.* New York: Zone Books, Swerve Editions, 1997.

———. *War in the Age of Intelligent Machines.* New York: Zone Books, Swerve Editions, 1991.

Deneubourg, J. L., S. Aron, S. Goss, J. M. Pasteels, and G. Duerinck. "Random Behavior, Amplification Processes and Number of Participants: How They Contribute to the Foraging Properties of Ants." *Physica D* 22 (1986): 176–86.

Dennett, Daniel C. *Brainchildren: Essays on Designing Minds.* Cambridge, Mass.: MIT Press, 1998.

———. *Consciousness Explained.* Boston, London, and Toronto: Little Brown, 1991.

De Waal, Franz. *Chimpanzee Politics.* Baltimore, Md.: Johns Hopkins University Press, 1982.

Diamond, Jared. *Guns, Germs and Steel: The Fates of Human Societies.* New York and London: W. W. Norton, 1997.

———. *Why Is Sex Fun?: The Evolution of Human Sexuality.* New York: Basic Books, 1997.

Dickens, Charles. *Bleak House.* Boston: Houghton-Mifflin, Riverside Editions, 1956.

Donaldson, Margaret. *Children's Minds.* New York: W. W. Norton, 1978.

Dyson, George B. *Darwin Among the Machines: The Evolution of Global Intelligence.* New York and Menlo Park, Calif.: Addison Wesley, 1997.

Edelman, Gerald M. *Bright Air, Brilliant Fire: On the Matter of Mind.* New York: Basic Books, 1992.

———. "Building a Picture of the Brain." *Daedalus* 127 (Spring 1998): 37–69.

———. *Topobiology: An Introduction to Molecular Embryology.* New York: Basic Books, 1988.

Edelman, Gerald, and Giulio Tononi. *A Universe of Consciousness: How Matter Becomes Imagination.* New York: Basic Books, 2000.

Editors of *Scientific American. The Scientific American Book of the Brain.* New York: Lyons Press, 1999.

Engels, Friedrich. *The Condition of the Working Class in England.* New York: Penguin, 1987.

Etcoff, Nancy. *Survival of the Prettiest: The Science of Beauty.* New York and London: Doubleday, 1999.

Fallows, James. *Breaking the News: How the Media Undermine American Democracy.* New York: Vintage, 1997.

Flake, Gary William. *The Computational Beauty of Nature: Computer Explorations of Fractals, Chaos, Complex Systems, and Adaptation.* Cambridge, Mass.: MIT Press, 1998.

Fox Keller, Evelyn. "The Force of the Pacemaker Concept in Theories of Aggregation in Cellular Slime Mold." In *Reflections on Gender and Science.* New Haven, Conn.: Yale University Press, 1996.

Garfinkel, Alan. "The Slime Mold Dictyostelium as a Model of Self-Organization in Social Systems." In *Self-Organizing Systems: The Emergence of Order,* edited by F. Eugene Yates. New York and London: Plenum Press, 1987.

Garreau, Joel. *Edge City: Life on the Frontier.* New York: Doubleday, 1991.

Gelerntner, David. *Mirror Worlds, or the Day Software Puts the Universe in a Shoebox . . . How It Will Happen and What It Will Mean.* New York and Oxford: Oxford University Press, 1992.

Gell-Mann, Murray. *The Quark and the Jaguar: Adventures in the Simple and the Complex.* New York: W. H. Freeman and Co., 1994.

Gies, Joseph, and Frances Gies. *Life in a Medieval City.* New York: Harper Perennial, 1969.

Gleick, James. *Chaos: Making a New Science.* New York and London: Penguin Books, 1987.

Gordon, Deborah. *Ants at Work: How an Insect Society Is Organized.* New York: Free Press, 1999.

Gould, Stephen Jay. *Full House: The Spread of Excellence from Plato to Darwin.* New York: Harmony Books, 1996.

———. *The Panda's Thumb: More Reflections on Natural History.* New York and London: W.W. Norton, 1980.

———. *Wonderful Life: The Burgess Shale and the Nature of History.* New York and London: W. W. Norton, 1989.

Guzeldere, Guven, and Stefano Franchi, eds. "Bridging the Gap: Where Cognitive Science Meets Literary Criticism." *Stanford Humanities Review* 4, no. 1 (Spring 1994): suppl.

Hall, Peter. *Cities of Tomorrow: An Intellectual History of Urban Planning and Design in the Twentieth Century.* Cambridge, Mass., and Oxford, UK: Blackwell Publishers, 1988.

Harvey, David. *The Urban Experience.* Baltimore, Md.: Johns Hopkins University Press, 1989.

Hayles, N. Katherine. *Chaos Bound: Orderly Disorder in Contemporary Literature and Science.* Ithaca, N.Y., and London: Cornell University Press, 1990.

———. *Chaos and Order: Complex Dynamics in Literature and Science.* Chicago: University of Chicago Press, 1991.

Hibbert, Christopher. *Florence: The Biography of a City.* New York and London: W. W. Norton, 1993.

———. *The House of the Medici: Its Rise and Fall.* New York: William Morrow, 1975.

Hillis, Daniel. *The Pattern on the Stone: The Simple Ideas That Make Computers Work.* New York: Basic Books, 1998.

Hodges, Andrew. *Alan Turing: The Enigma.* New York: Walker and Co., 2000.

Hoffman, Donald D. *Visual Intelligence: How We Create What We See.* New York and London: W. W. Norton, 1998.

Hofstadter, Douglas. *Gödel, Escher, Bach: An Eternal Golden Braid.* New York: Basic Books, 1979.

———. *Le Ton beau de Marot: In Praise of the Music of Language.* New York: Basic Books, 1997.

Holland, John H. *Emergence: From Chaos to Order.* Reading, Mass.: Helix, 1998.

———. *Hidden Order.* Reading, Mass.: Helix, 1995.

Humphrey, Nicholas. *A History of the Mind: Evolution and the Birth of Human Consciousness.* New York: Springer-Verlag, Copernicus Editions, 1992.

Huxley, Aldous. *The Doors of Perception* and *Heaven and Hell.* New York: Harper Perennial, 1963.

Iberall, Arthur S. "A Physics for Studies of Civilization." In *Self-Organizing Systems: The Emergence of Order,* edited by Eugene F. Yates, ed. New York and London: Plenum Press, 1987.

Ince, D. C., ed. *Mechanical Intelligence.* Vol. 3 of *The Collected Works of Alan Turing.* New York: Elsevier Science Publishing, 1992.

Innes, Judith E., and David E. Booher. "Metropolitan Development as a Complex System: A New Approach to Sustainability." *Economic Development Quarterly* 13 (1999): 141–56.

Jacobs, Jane. *The Death and Life of Great American Cities.* New York: Vintage, 1961.

———. *The Nature of Economies.* New York: Modern Library Original, 2000.

Johnson, George. *Fire in the Mind: Science, Faith and the Search for Order.* New York: Vintage, 1995.

Johnson, Steven. *Interface Culture: How New Technology Transforms the Way We Create and Communicate.* San Francisco: Harper Edge, 1997.

Jones, Steve. *The Language of Genes: Solving the Mysteries of Our Genetic Past, Present and Future.* New York and London: Anchor Books, Doubleday, 1993.

Joyce, James. *Ulysses.* New York: Vintage, 1986.

Karsai, Istvan, and John W. Wenzel. "Productivity, individual-level and colony-level flexibility, and organization of work as consequences of colony size."

Proceedings of the National Academy of Sciences of the United States of America 95 (1998): 8665–69.

Katz, Peter. *The New Urbanism: Toward an Architecture of Community.* New York, San Francisco, and Washington, D.C.: McGraw-Hill, 1994.

Kauffman, Stuart. *At Home in the Universe: The Search for the Laws of Self-Organization and Complexity.* New York and Oxford: Oxford University Press, 1995.

Kelly, Kevin. *Out of Control.* New York: Addison-Wesley, 1994.

Kelso, J. A. Scott. *Dynamic Patterns: The Self-Organization of Brain and Behavior.* London and Cambridge, Mass.: MIT Press, 1999.

Kessin, Richard H., and Michiel M. Campagne. "The Development of a Social Amoeba." *American Scientist* 80, no. 6 (1992): 556–65.

Klein, Naomi. "The Vision Thing." *The Nation,* July 10, 2000.

Koch, Christof, and Gilles Laurent. "Complexity and the Nervous System." *Science* 284 (1999): 96–98.

Kowalewski, Stephen A. "The Evolution of Complexity in the Valley of Oaxaca." *Annual Review of Anthropology* 19 (1990): 39–58.

Kramer, Peter. *Listening to Prozac.* New York and London: Penguin Books, 1993.

Krugman, Paul. *The Self-Organizing Economy.* Oxford, UK, and Malden, Mass.: Blackwell Publishers, 1996.

Kuhn, Thomas. *The Structure of Scientific Revolutions,* 2nd ed. Chicago: University of Chicago Press, 1962, 1970.

Kunstler, James Howard. *The Geography of Nowhere: The Rise and Decline of America's Man-Made Landscape.* New York: Touchstone, 1993.

———. *Home from Nowhere: Remaking Our Everyday World for the 21st Century.* New York: Touchstone, 1996.

Kurzweil, Ray. *The Age of Spiritual Machines: When Computers Exceed Human Intelligence.* New York: Penguin Books, 1999.

Langton, Christopher, et al., eds. *Artificial Life II.* Redwood City and Menlo Park, Calif.: Addison Wesley, 1990.

Leonard, Andrew. *Bots: The Origin of New Species.* San Francisco: Hardwired Books, 1997.

Lessig, Lawrence. *Code, and Other Laws of Cyberspace.* New York: Basic Books, 1999.

Levy, Steven. *Artificial Life: A Report from the Frontier Where Computers Meet Biology.* New York: Vintage, 1992.

Lewin, Roger. *Complexity: Life and the Edge of Chaos.* New York: Macmillan, 1992.

Lewis, R. W. B. *The City of Florence: Historical Vistas and Personal Sightings.* New York: Farrar, Straus and Giroux, 1995.

Lippmann, Walter. *Public Opinion.* New York: Free Press, 1922, 1949, 1997.

Lucas-Dubredon, J. *Daily Life in Florence: In the Time of the Medicis.* New York: Macmillan, 1961.

Lumer, E. D., G. M. Edelman, and G. Tononi. "Neural dynamics in a model of the thalamocortical system. I. Layers, loops and the emergence of fast synchronous rhythms." *Cerebral Cortex* 7 (1997): 207–27.

Lynch, Kevin. *The Image of the City.* Cambridge, Mass.: MIT Press, 1997.

Macleod, Katrina, Alex Becker, and Gilles Laurent. "Who reads temporal information contained across synchronized and oscillatory spike trains?" *Nature* 395 (1998): 693–98.

Marcus, Steven. *Engels, Manchester, and the Working Class.* New York: W. W. Norton, 1974.

McCann, Kevin, Alan Hastings, and Gary R. Huxel. "Weak trophic interactions and the balance of nature." *Nature* 395 (1998): 794–98.

McIntosh, Anthony Randal, M. Natasha Rajah, and Nancy J. Lobaugh. "Interactions of Prefrontal Cortex in Relation to Awareness in Sensory Learning." *Science* 28 (May 1999): 1531–33.

McLuhan, Marshall. *Understanding Media: The Extensions of Man.* Cambridge, Mass., and London: MIT Press, 1994.

Minsky, Marvin. *The Society of Mind.* New York: Touchstone, 1985.

Mitchell, William J. *City of Bits: Space, Place and the Infobahn.* London and Cambridge, Mass.: MIT Press, 1995.

Mithen, Steven. *The Prehistory of Mind: The Cognitive Origins of Art, Religion and Science.* London: Thames and Hudson, 1996.

Moretti, Franco. *Atlas of the European Novel, 1800–1900.* New York and London: Verso, 1998.

Morgan, Gareth. *Images of Organization.* San Francisco: Berren-Koehler Publishers, 1997; and Thousand Oaks, Calif.: Sage Publications, 1997.

Morrison, Philip, Phylis Morrison, and the Office of Charles & Ray Eames. *Powers of Ten: About the Relative Size of Things in the Universe.* New York: Scientific American Library, 1999.

Mountcastle, Vernon B. "Brain Science at the Century's Ebb." *Daedalus* 127 (Spring 1998): 1–36.

Mumford, Lewis. *The City in History: Its Origins, Its Transformations and Its Prospects.* New York and London: Harcourt, Brace, Jovanovich, 1961.

———. "The Sky Line." *The New Yorker,* December 1, 1962, 148–77.

Murphy, Michael P., and Luke A. J. O'Neill, eds. *What Is Life? The Next Fifty Years: Speculations on the Future of Biology.* Cambridge, UK: Cambridge University Press, 1995.

Ornstein, Robert, and Richard Thompson. *The Amazing Brain.* Boston: Houghton-Mifflin, 1984.

Penrose, Roger. *The Emperor's New Mind: Concerning Computers, Minds, and the Laws of Physics.* New York: Penguin Books, 1991.

Philips, William A., and Wolf Singer. "In search of common foundations for cortical computation." *Behavioral and Brain Sciences* 20 (1997): 657–722.

Pinker, Steven. *How the Mind Works.* New York: Norton, 1997.

———. *The Language Instinct: How the Mind Creates Language.* New York: Harper Perennial, 1994.

Prigogine, Ilya, and G. Nicolis. *Exploring Complexity.* New York: W. H. Freeman, 1989.

Resnick, Mitchell. *Turtles, Termites, and Traffic Jams: Explorations in Massively Parallel Microworlds.* Cambridge, Mass., and London: MIT Press, 1999.

Restak, Richard. *Brainscapes: An Introduction to What Neuroscience Has Learned About the Structure, Function, and Abilities of the Brain.* New York: Hyperion, 1995.

Ridley, Matt. *Genome: The Autobiography of a Species in 23 Chapters.* New York: HarperCollins, 1999.

Rogers, Richard. *Cities for a Small Planet.* London: Faber and Faber, 1997.

Rosenstiel, Tom. *Strange Bedfellows: How Television and the Presidential Candidates Changed American Politics, 1992.* New York: Hyperion, 1993.

Rushkoff, Douglas. *Coercion: Why We Listen to What "They" Say.* New York: Riverhead Books, 1999.

———. *Media Virus!: Hidden Agendas in Popular Culture.* New York: Ballantine Books, 1994.

Rybczynski, Witold. *A Clearing in the Distance: Frederick Law Olmsted and America in the Nineteenth Century.* New York: Scribner, 1999.

Sacks, Oliver. *An Anthropologist on Mars.* New York: Vintage Books, 1995.

Sadler, Simon. *The Situationist City.* Cambridge, Mass., and London: MIT Press, 1998.

Schacter, Daniel L. *Searching for Memory: The Brain, the Mind and the Past.* New York: Basic Books, 1997.

Schelling, Thomas. *Micromotives and Macrobehavior.* New York and London: W. W. Norton, 1978.

Schreiber, Darren. "The Emergence of Parties: An Agent-Based Model." Online posting. www.swarm.org/community-links.html. March 20, 2000.

Schroeder, Manfred. *Fractals, Chaos, Power Laws: Minutes from an Infinite Paradise.* New York: W. H. Freeman and Co., 1991.

Selfridge, O. G. "Pandemonium: A Paradigm for Learning." In *Mechanization of Thought Processes. Proceedings of a Symposium Held at the National Physical Laboratory in November 1958.* London: Her Majesty's Stationery Office, 1959.

Selfridge, Oliver G., Edwina L. Rissland, and Michael A. Arbib, eds. *Adaptive*

Control of Ill-Defined Systems. New York and London: Plenum Press, 1984.

Sennett, Richard. *The Conscience of the Eye: The Design and Social Life of Cities.* New York: Knopf, 1990.

———. *The Fall of Public Man: On the Social Psychology of Capitalism.* New York: Vintage, 1978.

———, ed. *Classic Essays on the Culture of Cities.* Englewood Cliffs, N.J.: Prentice-Hall, 1969.

Shah, A. M., B. S. Baviskar, and E. A. Ramaswamy. *Complex Organizations and Urban Communities.* Vol. 3 of *Social Structure and Change.* New Dehli, London, and Thousand Oaks, Calif.: Sage Publications, Inc., 1996.

Shannon, Claude E. *The Mathematical Theory of Communication.* Chicago: University of Illinois Press, 1998.

Shapiro, Andrew L. *The Control Revolution: How the Internet Is Putting Individuals in Charge and Changing the World We Know.* New York: Century Foundation Books, 1999.

Stephenson, Neal. *Cryptonomicon.* New York: Avon Books, 1999.

Stopfer, Mark, Seetha Bhagavan, Brian H. Smith, and Gilles Laurent. "Impaired odor discrimination on desynchronization of odor-encoding neural assemblies." *Nature* 390 (1997): 70–74.

Storr, Anthony. *Music and the Mind.* New York: Free Press, 1992.

Strohman, Richard C. "The coming Kuhnian revolution in biology." *Nature Biotechnology* 15 (1997): 194–200.

Taylor, John. *The Race for Consciousness.* Cambridge, Mass., and London: MIT Press, 1999.

Thomas, Lewis. *The Lives of a Cell: Notes of a Biology Watcher.* New York: Penguin, 1974.

Thompson, E. P. *Customs in Common: Studies in Traditional Popular Culture.* New York: New Press, 1993.

Tononi, Giulio, and Gerald M. Edelman. "Consciousness and Complexity." *Science* 282 (1998): 1846–51.

———. "Consciousness and the Integration of Information in the Brain." *Consciousness: At the Frontiers of Neuroscience, Advances in Neurology* 77 (1998): 245–80.

Tononi, Giulio, Gerald M. Edelman, and Olaf Sporns. "A complexity measure for selective matching of signals by the brain." *Proceedings of the National Academy of Sciences of the United States of America* 93 (1996): 3422–27.

———. "Information in the brain." *Trends in Cognitive Sciences* 2 (1998).

———. "Measures of degeneracy and redundancy in biological networks." *Proceedings of the National Academy of Sciences of the United States of America* 96 (1999): 3257–62.

Tononi, Giulio, R. Srinivasan, D. P. Russell, and G. M. Edelman. "Investigating neural correlates of conscious perception by frequency-tagged neuromagnetic responses." *Proceedings of the National Academy of Sciences of the United States of America* 95 (1998): 3198–203.

Varela, Francisco, Evan Thompson, and Eleanor Rosch. *The Embodied Mind: Cognitive Science and Human Experience.* Cambridge, Mass., and London: MIT Press, 1993.

Waldrop, Mitchell M. *Complexity: The Emerging Science at the Edge of Order and Chaos.* New York and London: Simon and Schuster, 1992.

Watts, Duncan J., and Steven H. Strogatz. "Collective dynamics of small-world networks." *Nature* 393 (1998): 440–42.

Weaver, Warren. "Science and Complexity." *American Scientist* 36 (1948): 536.

Weiss, Michael J. *The Clustered World: How We Live, What We Buy, and What It All Means About Who We Are.* New York, London, and Boston: Little, Brown, 2000.

White, Leslie. "Energy and the Evolution of Culture." *American Anthropologist* 45 (1943): 335–56.

Wiener, Norbert. *Cybernetics: or, Control and Communication in the Animal and the Machine.* Cambridge, Mass.: MIT Press, 1948.

Williams, Raymond. *The Country and the City.* New York: Oxford University Press, 1973.

Willis, William, and Robert Grossman, eds. *Medical Neurobiology.* St. Louis: CB Moseby Company, 1977.

Wilson, Edward O. *Consilience: The Unity of Knowledge.* New York: Knopf, 1998.

———. *Sociobiology.* Abr. Ed. Cambridge, Mass., and London: Harvard University Press, 1975, 1980.

Wilson, Edward O., and Bert Holldobler. *The Ants.* Cambridge, Mass.: Harvard University Press, 1990.

Wright, Robert. *NonZero: The Logic of Human Destiny.* New York: Pantheon Books, 2000.

ACKNOWLEDGMENTS

In a way, the idea for this book began with a thirtieth-birthday present—actually, two separate presents that by coincidence happened to be the same book, one from my father and one from my oldest friend, the architect and Web designer Eric Liftin. The book in question was a large-format atlas of nineteenth-century city maps, and it included a striking image of Hamburg that looked uncannily like a profile view of a human brain. I had been following two separate tracks of reading over the preceding year—one on cities and the other on the mind. Somehow that image of Hamburg triggered a vague connection in my head, and I began to wonder if perhaps the two paths harbored a secret intersection.

Other roots should be mentioned. I'd written about complexity theory and culture for a 1996 *Lingua Franca* essay. A few pages of this book originate with a 1998 *Harper's* piece on pattern-matching and the Alexa software, an essay that itself developed out of the "Agents" chapter from my last book, *Interface Culture.* In early 1999, I wrote an introduction to *ID Magazine*'s year-end-awards issue that addressed control in interactive design and video games. *Brill's Content* was kind enough to let me ruminate on Robert Wright's "global brain" idea for a few thousand words, as I was in the middle of finishing the manuscript. And much of this book connects with something I wrote or edited for *FEED* over the past few years—particularly our special issues on the brain, video games, and cities. So a special thanks to all my editors over the past six years: my coeditor-in-chief at FEED, Stefanie Syman; Alex Star; Sam Lipsyte; Amanda Griscom; Austin Bunn; James Ryerson; Alex Abramovich; Ben Cosgrove; Deborah Shapiro; Elaine Blair; Christiane Culhane; Mark Van de Walle; David Kuhn; Susan Burton; Franco Moretti; and Chee Perlman. Stefanie, Jamie, and Eric were also kind enough to read early versions of this manuscript. (My colleague Matt Goldberg read the manuscript by osmosis.) I'm grateful to them for their comments and suggestions. They are, of course, responsible for any errors, and all the good parts are mine.

I was afforded a unique opportunity during the writing of this book, in that parts of my day job at Automatic Media involved helping with the design and implementation of self-organizing software: mainly in our Plastic.com site, which was built on the Slashdot code. It's not often that a writer gets to build something as he or she is writing about it, and it's equally unusual to get to do that building in the company of so many bright minds. So, special props to Lee deBoer, Joey Anuff, Matt Goldberg, Michael Kolbrener, Freyja Balmer, Jon Phelps, Rob Francis, and J. J. Gifford. They deserve extra credit for suffering through all my overcaffeinated riffs on clusters and pointer nodes.

This book was greatly enhanced by interviews I conducted with Manuel De Landa, Richard Rogers, Deborah Gordon, Rob Malda, Jeff Bates, Oliver Selfridge, Will Wright, David Jefferson, Evelyn Fox Keller, Rik Heywood, Mitch Resnick, Steven Pinker, Eric Zimmerman, Nate Oostendorp, Brewster Kahle, Andrew Shapiro, and Douglas Rushkoff. I recall more than a few casual conversations that also had an impact, primarily ones that involved David Shenk, Ruthie Rogers, Roo Rogers, Mitch Kapor, Kevin Kelly, Annie Keating, Nicholas Butterworth, Kim Hawkins, Rory Kennedy, Mark Bailey, Frank Rich, Denise Caruso, Liz Garbus, Dan Cogan, Penny Lewis, John Brockman, Rufus Griscom, Jay Haynes, Betsey Schmidt, Stephen Green, Esther Dyson, and my students at NYU's ITP program, where Red Burns generously invited me to teach a graduate seminar on emergent software. My family, as always, was a constant source of ideas and encouragement—particularly my two direct connections to the world of medicine, my mother and my sister Sallie.

For most of the writing of this book, Andrew Shroeder played an invaluable role as research assistant, tracking down obscure essays and reading alongside me. (Jay Demas and Josh Saunders also helped with important research along the way.) My agent, Lydia Wills, once again did a masterful job of nudging an unwieldy first-draft proposal toward something that could actually be published. My editor at Penguin UK, Stefan McGrath, made a number of timely and astute contributions to the draft manuscript. At Scribner, Rachel Sussman was incredibly patient with my late arrivals. As for my gifted editor Gillian Blake—not only did she not flinch when I told her about my idea of opening the whole book with slime molds, she also provided exactly the conceptual and sentence-by-sentence guidance that I needed in putting together a complicated, multithreaded book.

Then there's my wife, Alexa Robinson. There is no finer line editor in the land, and no better advocate, sounding board, and support system. She is, in more ways than one, my ideal reader. This book—along with our marriage—turns out to be one of those future collaborations I alluded to in the last acknowledgments, but I'm fairly sure there are more to come.

Nearly four years ago, days after Alexa and I moved into our apartment in the West Village, I finally got around to reading Jane Jacobs's *Death and Life of the Great American Cities.* I knew Jacobs had lived in the Village while writing the book, but I didn't know the exact whereabouts. From the very first chapter it was clear that she must have lived somewhere nearby. About a hundred pages in, with the help of the Web, I tracked down her actual residence: no more than three blocks from our apartment. All through the writing of this book, I could see the roof of Jacobs's old building from the study I was working in. I could see the rooftops and the sidewalks of the whole West Village sprawled out below me, the urban ballet that Jacobs had written about so powerfully forty years before. If books like this one require acknowledgments, they have to start—or end—with that great, shifting energy and its connective powers. This is a city book, both in subject matter and in inspiration. If you're reading these words in a comparably thriving city, put the book down, step outside into the roaring streets, and make your own connections.

March 2001
New York City

INDEX